解码智能时代

从中国国际智能产业博览会
瞭望全球智能产业（2018—2022）

黄桷树财经　编著

撰稿：张山斯　胡　浩　何永红　李全龙　吴荣飞
　　　甘　勤　杨　勤　陈　思　李雪妍

翻译：王　蓉　胡文江　程　锐　吴　链　周卫璇
　　　余　娇　邓琳姿　陈　佳　柳林康　曾　杰

重庆大学出版社

图书在版编目（CIP）数据

解码智能时代：从中国国际智能产业博览会瞭望全球智能产业：2018—2022：汉、英／黄桷树财经编著；王蓉等译. -- 重庆：重庆大学出版社，2022.8（2023.4 重印）

书名原文：Decrypting the Intelligent Era: Overlook the Global Intelligent Industry from Smart China Expo（2018—2022）

ISBN 978-7-5689-3483-1

Ⅰ.①解… Ⅱ.①黄… ②王… Ⅲ.①人工智能 – 汉、英 Ⅳ.① TP18

中国版本图书馆 CIP 数据核字 (2022) 第 133699 号

解码智能时代：从中国国际智能产业博览会瞭望全球智能产业（2018—2022）

JIEMA ZHINENG SHIDAI: CONG ZHONGGUO GUOJI ZHINENG CHANYE BOLANHUI LIAOWANG QUANQIU ZHINENG CHANYE（2018—2022）

黄桷树财经　编著

王蓉　等译

策划编辑：雷少波　许璐　杨琪

责任编辑：许璐　杨琪　　　版式设计：许璐
责任校对：谢芳　　　　　　责任印制：张策

*

重庆大学出版社出版发行
出版人：饶帮华

社址：重庆市沙坪坝区大学城西路 21 号
邮编：401331
电话：（023）88617190　88617185（中小学）
传真：（023）88617186　88617166
网址：http://www.cqup.com.cn
邮箱：fxk@cqup.com.cn（营销中心）
全国新华书店经销
POD：重庆市圣立印刷有限公司

*

开本：720mm×960mm　1/16　印张：19.5　字数：347 千
2022 年 8 月第 1 版　　2023 年 4 月第 4 次印刷
ISBN 978-7-5689-3483-1　定价：88.00 元

世界正进入数字经济快速发展的时期，5G、人工智能、智慧城市等新技术、新业态、新平台蓬勃兴起，深刻影响全球科技创新、产业结构调整、经济社会发展。近年来，中国积极推进数字产业化、产业数字化，推动数字技术同经济社会发展深度融合。

　　在上海合作组织成立 20 周年之际，中国愿同各成员国弘扬"上海精神"，深度参与数字经济国际合作，让数字化、网络化、智能化为经济社会发展增添动力，开创数字经济合作新局面。

摘自新华社北京 2021 年 8 月 23 日电：《习近平向中国—上海合作组织数字经济产业论坛、2021 中国国际智能产业博览会致贺信》

目录

CONTENTS

第五章
重庆：一种全新智慧城市的发展探索

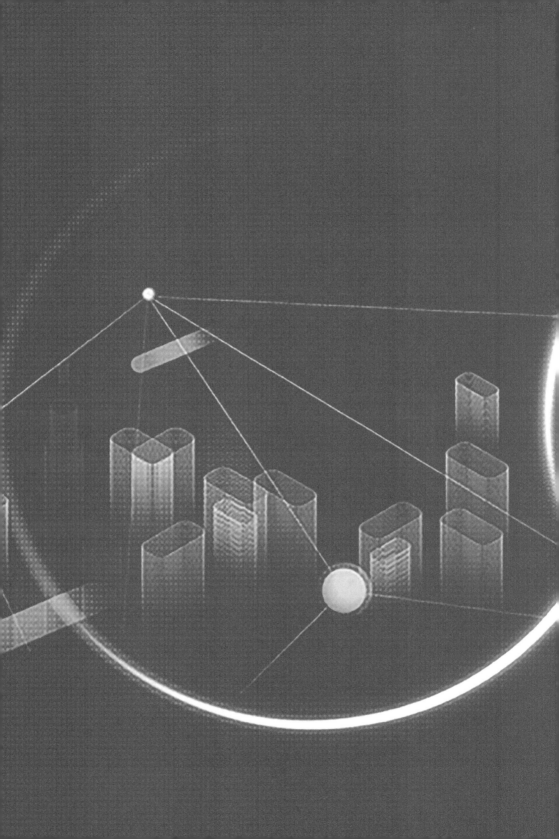

第一章

智博会五年：见证智能新时代的到来

忽然之间，全球科技界有些热闹，一种看不真切却又浓墨重彩的时代景象，一种听不清晰却又振聋发聩的时代声音，席卷而来，充斥周际。如旌旗蔽空，如鼓角齐鸣，如万马奔腾，如八音迭奏，让人浑身骤起一种莫名的亢奋，又于亢奋中毫不犹豫地投身其间。

过去五年，就是在这种轰轰烈烈的进程中，人工智能极速成为全球各个国家、各个城市、各家企业、各种专家面向未来世界的共同话题，而五年智博会，汇集了整个智能产业的力量与想象，也因此见证了一个智能新时代的发端与形成。

第一节
人工智能这五年，整个世界的进化缩影

站在 2022 年，回望 2018 年，每个人都会有不同的回忆。短短五年，弹指一瞬，而仔细对比其中的变化，又难免让人产生沧海桑田之感。

你还记得共享经济的如火如荼吗？共享单车、共享出行、共享民宿、共享充电宝、灵活用工……突然变得炙手可热，中国的共享经济因为广泛覆盖交通出行、房屋住宿、知识技能、生活服务等众多领域，迅速成为全球共享经济的创新者和引领者。

你还记得新零售的横空出世吗？依托于中国电子商务十年的蓬勃发展，以及互联网对传统线下商业的反向渗透，互联网零售的触角开始以各种创新的形式延伸到线下，无人商店、餐饮外卖、商超到家、社区团购、送药上门、互联网买菜……一场轰轰烈烈的中国新零售变革，因为丰富的想象、勇敢的创新、庞大的市场，已经成为全球数字经济的创新源头。

这些琳琅满目的中国新经济探索，从生涩走向成熟，从 2018 年起，中国创新终于开始在部分领域引领全球。

2010 年左右，中国的创新者把握住移动互联网的创新机遇，同步全球技术趋势，与全球创新者站在几乎相同的起点，而依托于国内蓬勃增长的市场需求，更独享一份特有的中国消费升级红利。在此期间，中国的创业者不再像互联网发展早期那样较多地复制国外成熟模式，而是越来越多地进行中国本土原创性的创新，并开始对外输出，将影响扩散到全球。

而本书所关注的重点正是从 2018 年开始，在中国创新风靡全球的大背景下，全球经济正酝酿着切换创新的主引擎，人工智能从实验室走向全球经济、生活与社会治理各个层面的全新智能时代到来的五年旅程。

时代造势神工天巧，城市顺势开物成务，让我们还是从 2018 年的起点开始说起吧。

中国创新，开始影响全球

2017 年 5 月，一位名叫彼得的罗马尼亚在华研究生，极大地激发了很多中国人的自豪感。他在接受媒体采访的时候，对着镜头谈起在中国生活的感受，脸上充满兴奋："我很痴迷中国古代的四大发明，现在中国发展得太快了，已经有了'新四大发明'。"

彼得口中的"新四大发明"，来自北京外国语大学丝绸之路研究院发起的一项调查。"一带一路"沿线 20 个国家的在华留学生们经过投票，评选出了"最想带回自己国家的中国生活方式"，排在前面的是：高速铁路、网络购物、移动支付和共享单车。

严格意义上来说，"新四大发明"从技术视角来看没有一项是中国原创的，但由于这四种创新在中国市场的应用，无论场景构建，还是模式创新，以及形成的应用规模，在全球都遥遥领先，因此被评为中国"新四大发明"，并不为过。

伴随着中国经济的飞速发展，又恰逢移动通信技术在全球范围内的极速变革，中国成为移动互联网创新的热土，在全球范围内，中国消费者也成为积极拥抱创新、体验创新、参与创新最为积极的群体。

先说移动支付，日本人首先发明了二维码，但把二维码用于移动

● 高速铁路被誉为中国"新四大发明"之一
High-speed railway is known as one of the "New Four Great Inventions" of China.

支付，并将其从一项技术推行到全社会的大规模应用，中国对全球市场可谓贡献巨大。

从 2013 年开始，历经网约车、共享单车几番竞争，中国移动支付的普及速度惊人，仅仅几年时间，连摆摊卖水果的阿姨都在用二维码收款。2016 年，中国移动支付市场规模达到 157.55 万亿元人民币，而据全球咨询公司 Forrester Research 估算，2016 年，美国的移动支付市场规模是 1120 亿美元，两者之间体量相差 200 倍。[1]

再说外卖送餐，美国最早的外卖平台 Doordash 成立于 2013 年，这马上启发了当时已成立 4 年的饿了么和成立了 3 年的美团，它们迅速切入餐饮外卖领域，几乎无时差地在中国市场展开了全面竞争。

到了 2017 年，中国的外卖市场已成燎原之势。在众多小平台被

1　李延霞，新华社，《央行报告：2016 年我国移动支付金额同比增长近五成》，2017 年 3 月 16 日。

案来自于创造者的玩笑，还是来自于人工智能的真实想法，显然都足以让所有人大惊失色。

2017 年，在大量的新技术、新硬件、新算法之中，人工智能得到广泛应用。而和 AlphaGo 挑战人类的围棋比赛相较，其实在大多数场景下，人工智能都不屑于与人类比赛，甚至是一登场，就直接否定了比赛的意义。

这一年的"双 11"电商狂欢中，阿里的 AI 设计师"鲁班"走马上任，号称"1 秒钟就能制作 8000 张海报"，出尽了风头。没有任何一位设计师收到这场比赛的通知，因为在这个量级的比赛上，没有任何人类拥有参赛的资格。

并不是所有的人工智能技术，都在 2017 年才完成自身的技术突破。但在这一年，人工智能的大规模应用，却以超高的频次、超多的场景、超级的震撼，面向公众上演了一出又一出前所未有的大戏。

2017 年，因此成为业界公认的人工智能应用元年。

震惊、悲伤、兴奋与担忧，构成了 2017 年人类面对人工智能时，颇为复杂的情绪。

每一个国家、每一座城市、每一家企业、每一位个人，显然到了一个应该郑重思考的关键时刻：如何迎接智能时代的到来？如何定位智能时代的自己？

第二节
畅想与实践的交替，连续五年持续互动

　　谈论方兴未艾的人工智能，有一个惯常的误会需要消除，那就是：在本质上，这是一种生产技术的进步，而不是生存文明的挑战。认同了这一点，对它的许多担忧可能会有所缓和，最起码不会影响行动的选择。

　　在全球范围内，各国政府和创新企业，面对智能时代的到来都远比霍金等物理学家更为乐观积极。

智能时代的考题，一份城市的答卷

　　在中国，早在 2016 年 3 月，"人工智能"就被写入《中华人民共和国国民经济和社会发展第十三个五年规划纲要》；而到了 2022 年，"人工智能"已经连续六年被写入国务院的政府工作报告。

　　其中最关键的时间点，仍然是 2017 年。中国从国家战略层面，将人工智能的发展隆重地提上了日程：2017 年 7 月 20 日，国务院发布《新一代人工智能发展规划》，提出了面向 2030 年我国新一代人工智能发展的指导思想、战略目标、重点任务和保障措施，部署构筑我国人工智能发展的先发优势，加快建设创新型国家和世界科技强国。

　　2017 年的这个发展规划，究竟有多么重要？

这个问题，需要站在更长久的历史维度，才能更客观地进行评价。但这丝毫不影响我们在五年之后的今天，感受高屋建瓴的国家战略对于智能产业的强大推动力，以及由此带来的深刻变化。

甚至会让人产生一种后怕，如果国家的发展规划来得晚一点，哪怕只是晚一年，将会多么被动。

2016 年至 2017 年，最早一批发布人工智能战略的国家分别是美国、中国、日本和英国。从 2017 年到 2018 年，全世界的主流大国都在跃跃欲试，着眼于面向下一个时代的战略竞争布局，核心就是人工智能产业。到了 2018 年，几乎全球所有主要经济体，如德国、法国、意大利、印度、韩国、俄罗斯、新加坡等都相继推出了自己的人工智能发展战略。

像这样，世界上近乎所有国家，在接近的时间点上集体发布属于自己国家的产业发展战略，在历史上似乎还是首次。即便是涉及国家核心信息基础建设的 5G 战略，各个国家的战略规划动作也没有这么整齐。

一场围绕人工智能的国家竞赛，在 2017 年启动了。

五年之后，据 IDC 最新发布的数据，全球人工智能（AI）市场将增长近五分之一，2022 年市场规模将达到 4328 亿美元，预计今年人工智能支出将增长 19.6%，其中包括硬件、软件和服务。[1]

回到 2017 年的中国，在国务院发展规划的刺激效应下，企业界与创投圈围绕人工智能的投融资，展现出空前的繁荣景象。企查查数据显示，人工智能赛道在 2016 年至 2018 年的融资事件数量持续保持在 900 件以上，2019 年后市场有一定的理性回归，但仍在 500 件以上。

在发展规划的指引下，中国不同省市之间也开始了人工智能发展

1　拾壹，人民邮电报，《IDC：2022 年全球 AI 市场规模达到 4328 亿美元　增长近 20%》，2022 年 3 月 9 日。

竞赛，纷纷从自身的角度开始智能产业的深入布局。

然而智能产业毕竟是一个几乎涉及全社会所有经济、生活和公共治理的庞大产业，每个省市的产业基础、社会资源、人才构成完全不同。该以什么样的姿态、从哪个角度去拥抱智能产业，成为摆在各个省市面前的限时必答题。

答得好，就能抓住时代机遇，改变城市命运；答不好，就有可能错过机会，被动面对未来。

在《新一代人工智能发展规划》发布之后，反应最快的福州、贵阳、重庆和上海第一时间分别以"数字化""大数据""智能产业""人工智能"为关键词，于 2018 年申请举办了国家多部委联合主办支持的国际顶级智能产业峰会：数字中国建设峰会、中国国际大数据产业博览会、中国国际智能产业博览会和世界人工智能大会。

之后，以人工智能、数字化、大数据、云计算等智能产业关键词

● 2021 智博会展览现场
Smart China Expo 2021 site

为主题的峰会、展会、论坛、博览会，如雨后春笋般在全国各大城市出现。

值得一提的是，在 2018 年的四大智能产业峰会之中，有三个都是首次申请、首届举办，而在贵阳举办的"2018 中国国际大数据产业博览会"，其实已经是第二届。早在 2017 年 5 月，国务院《新一代人工智能发展规划》出台的两个月前，贵阳就已经成功举办了首届"数博会"，对一个大时代的到来，体现出了一种敏锐的先知先觉。

无论如何，时代的必然与城市的偶然，共同决定着重庆这座制造业之城，面向未来智能产业的关键答卷。

四届智博会，五年变形记

每年七八月份，重庆的夏天持续高温，按照过往的经验，重庆市民会选择外出避暑，但最近几年，很多人调整了这一生活习惯。

从 2018 年起，智博会永久落户重庆，每年 8 月份，智博会准时举办，这已经成为很多人新的夏日期待、新的习惯。

来自全球各地的智能产业科学家、企业家、创新者、研究者，汇集在火热的重庆，出席火热的智博会。他们围绕人工智能、数字经济、智能制造等专题，向全球传递了最前沿的创新理论、创新知识、创新技能、创新经验、创新模式，碰撞出新智慧，贡献出新成果。

既然是最前沿的，当然要是最火热的，对于重庆而言，这既是火炉城市的原始温度，也是智慧名城的未来愿景。

对于一座城市而言，一个为期只有几天的活动，到底能有多重要？或许很多读者会发出类似的疑问。

让我们先暂时把各种嘉宾演讲、论坛对话、产品展示、互动大屏放在一边，先看一组数据：

● 在 2021 智博会上，重庆市璧山区举行基于量子密钥分发（QKD）的应用技术发布与签约仪式

At the 2021 Smart China Expo, the release and signing ceremony of application technology based on Quantum Key Distribution (QKD) was held in Bishan District, Chongqing.

2018 年，智博会签约重大项目共 501 个；

2019 年，智博会签约重大项目共 530 个；

2020 年，受疫情影响，智博会主要采取线上举办模式，很多企业与项目方无法到现场，即便如此也签约了重大项目共 71 个；

2021 年，智博会签约重大项目 92 个，虽然整体规模与 2020 年相当，但 10 亿以上、50 亿以上项目均大幅增加，单一项目体量变大了。

这笔流水账，官方读得很铿锵，媒体写得很豪迈，市民看得很兴奋。

思想的交流和项目的签约是连在一起的，这是双重的火花。

解码智能时代 从中国国际智能产业博览会瞭望全球智能产业（2018—2022）

思想的交流，让重庆打开了脑门，去了解世界，去仔细观察正在变化的智能时代；而项目的签约，则使重庆打开了城门，去迎接客人，去联手创造深耕重庆的智能产业。

借助智博会，无论是在国内，还是在国际，重庆这座中国西部城市的知名度都在快速提升。而面向全球优质企业、优秀人才，重庆更是及时释放出智能时代创新之城的活力，吸引他们到重庆来创业。

作为 2019 智博会落地项目，"重庆工业大数据创新中心"两年来通过推动互联网、大数据等信息技术与实体经济深度融合，帮助了多家渝企实现降本增效。目前，该创新中心已在全市建成 8 家智能化改造示范工厂，全年累计将为 7000 多家中小企业提供"上云"服务。

2021 年 8 月 5 日，重庆市经信委透露，当年上半年重庆新增 1.6 万家企业"上云上平台"，目前全市已有 7 万多家企业迈上"云端"。[1]

2019 年、2020 年、2021 年，重庆市技术创新示范企业名单中，累计共有 125 家企业入选。

类似的数据，如果要枚举，还有很多很多维度；类似的突破，如果要罗列，还有很多很多捷报。在这些数据背后，相信读者能感受到，一座城市在智能时代焕发的生机，思维在转变，技术在迭代，产业在成长，捷报在频传，重庆企业技术创新突破正在加速涌现。

为这些数字进行溯源归因，很显然，最终的答案都会指向智博会的开端与持续。

---

1 夏元，重庆日报，《智博会落地项目助力工业互联网发展提速 7 万多家渝企迈上"云端"》，2021 年 8 月 6 日。

智博会大舞台不断变化的"C位"

从 2018 年到 2022 年，智博会舞台上"C位"的变化，同样体现出了一种时代的变迁与一座城市的进化。

在 2018 智博会上，除了各国政要、学术专家，站在舞台上为一个全新时代的到来而呐喊的演讲者，就是清一色的互联网企业家。他们在互联网的数字世界中冲浪多年，对于汹涌而来的数字化与智能化浪潮，拥有天生敏锐的嗅觉和积极拥抱的豪情，也更多地在演讲内容中体现了对于未来的豪情壮志。

而到了 2021 年，智博会上数字化与智能化的氛围愈加浓郁，而发表演讲的嘉宾中，来自传统制造业的企业家越来越多。在他们的分享中，数字化与智能化司空见惯，甚至绝大部分内容都不再是想象与预言，而是充满了落地的技术、实践的方案、取得的成果和沉淀的数据。

在这种现象背后，实体产业已经成为人工智能"大展拳脚"的舞台。事实上，在过去五年中，智能产业的绝大部分成就都来源于传统制造型企业与数字化创新型平台的广泛合作，这说明中国互联网产业与传统制造业顺畅地完成了数字化、智能化的衔接与融合。

为了深度参与数字经济国际合作、不断扩大对外交流合作，2021 智博会和中国—上海合作组织数字经济产业论坛同期举办。上合组织秘书长诺罗夫对中国数字经济的发展速度印象尤为深刻，他在致辞中提到："《中国数字经济发展白皮书（2021）》显示，中国数字经济规模居全球第二，中国数字经济的增长速度是 GDP 的三倍以上，这也说明数字经济在推动经济发展中扮演着关键角色。2020 年，中国数字经济规模达到 39.2 万亿元（约 6 万亿美元）占 GDP 的 38.6%，有效支持疫情防控工作和国家经济发展。"[1]

---

1 苏晓，人民邮电报，《〈中国数字经济发展白皮书〉发布，2020 年我国数字经济规模达到 39.2 万亿元》，2021 年 5 月 6 日。

● 2021 智博会上，中国—上合组织数字经济合作展馆

China-SCO Digital Economy Cooperation pavilion at the 2021 Smart China Expo

　　说到底，人工智能的技术和理念，需要转换为数字经济的生产力，才符合智能产业时代的大趋势。

　　永久落户重庆的智博会舞台上的变化，恰好又是重庆乃至整个中国经济结构的缩影。

第三节
思维与产品的创新，连续五年恰逢其会

智博会的连续举办，在很大程度上改变了重庆市与这个世界的互动方式，在瞬息万变的智能产业时代，重庆市的发展思路异常清晰，入驻重庆的企业家们也表现得非常果决。

纵观古今中外知名城市的崛起，不难发现，在一个风云代际的历史关口，只有一座城市的顶层设计与底层奋斗达成关键共识，城市才能在面向未来的成长中焕发新生。所以，当重庆这座城市自上而下所有的人都清晰地看到了智能时代赋予的崛起机会时，整座城市自然就开始奋勇向前。

对于重庆来说，举办智博会不仅是要把一场会办好，而且是通过智博会完成一座城市智能产业基因的再造：要升级制造，就连接工厂设备；要改善生活，就创新智能生活；要优化治理，就打通街头数据。连接了、创新了、打通了，才能坐下来聊聊经验、谈谈收获、看看沉淀，再抬头、再起步，风景已经完全不同。

仙桃数据谷，从一片荒山到一块热土

2014年4月，重庆市公布了一个全新的科技产业园规划。规划范围总面积为2674亩，选址在重庆市渝北区双龙湖街道仙桃村。对

解码智能时代　从中国国际智能产业博览会瞭望全球智能产业（2018—2022）

于这个位置，重庆市民普遍感叹"太远了"。

到底有多偏呢？

当时，在重庆市民心目中，刚建成不久的重庆园博园离主城区已经够远了，而这个地方还在园博园正北方向 6 公里处。

规划之初，这个地方还是荒山野坡，所见之处杂草丛生、乱石嶙峋，没有一条像样的道路，没有一个明显的标识，没有完善的电力设施，没有统一的供排水系统。

现在，逐步落成的重庆仙桃数据谷，随着"创新谷""智慧谷""生态谷"的加快建设，已发展成为大数据产业生态谷，并获得了"中国最具活力软件园"等诸多殊荣。[1]

乘着智能时代到来的东风，这一片荒山到一块热土的转变，仅仅用了八年时间。

仙桃数据谷到底有多热门呢？

这里已经成为全国智能车载软件企业第二总部的首选目的地。重庆市汽车软件产业基地在此正式落户，已聚集长安软件、中科创达、创通联达、黑芝麻智能科技等智能汽车产业重点企业十余家，相关从业人员近千人。已建成全国首个 5G 自动驾驶开放道路场景示范运营基地，可在 5 千米长的循环路线上进行智能网联汽车的示范运营。

8 年过后，这里不仅有造型独特的"小蛮腰"、百米高连廊"指环王"、"全玻璃立方"的会议中心、积木造型的大数据学院等科技感十足的楼宇集群，在招商引资与产业孵化上，更是成绩斐然：截至 2022 年 3 月，仙桃数据谷累计注册企业 1293 家，入驻企业 250 家，申请知识产权 1749 件、授权 1232 件，版权登记量、专利软著量等知识产权成果实现稳步增长。[2]

这就是重庆仙桃数据谷，目前已经成为全球智能产业炙手可热、

---

1 张亦筑，重庆日报，《仙桃数据谷，魅力大在哪》，2021 年 8 月 5 日。
2 梁浩楠，《华龙网》，《重庆唯一！仙桃数据谷入选全国版权示范园区》，2022 年 3 月 19 日。

● 重庆仙桃数据谷，已经成为全球智能产业争相入驻的目的地
Chongqing Xiantao Big Data Valley has become a hot destination for global smart industries to compete for residence.

争相入驻的目的地。

从一片荒芜到一块热土，八年前后的对比，同一个仙桃，判若两谷。仙桃数据谷的八年蜕变史，恰恰也是重庆智能产业的一个缩影。

智博会，重庆市的"第二招商局"

仙桃数据谷，只是重庆市挺进智能时代的一个缩影。而五年以来，类似仙桃数据谷的蜕变，其实在这座城市的各个角落同步发生。

或许最初，对于很多出席 2018 智博会的嘉宾而言，只是将这当成一场普通的峰会。飞到重庆，出席活动，发表演讲，然后就可以打

● 2021 智博会上，腾讯与吉利工业互联网平台战略合作签约仪式

At the 2021 Smart China Expo, Tencent and Geely Industrial Internet Strategic Cooperation Signing Ceremony

道回府，回到自己企业所在的地方，继续探索智能时代的未来。

后来，智博会成为很多人进入重庆的一扇门。而对于重庆市而言，智博会也成了重庆市硕果累累的"第二招商局"。

2018 年，因为智博会，腾讯、阿里、华为、百度、浪潮、中国电子、科大讯飞等国内科技龙头企业相继落地。它们或是在重庆成立西南总部，或是在重庆打造西南地区客户体验和数据中心，开始在重庆开展人工智能的更深度探索。

此后三年，更是有无数企业通过智博会的舞台走进重庆的智能产业生态圈，特别是在智能化、大数据、高端电子信息、新材料等领域，名企荟萃、络绎不绝。

2021 年 8 月 23 日，李书福现身重庆并出席智博会的消息迅速登上了国内各大媒体。这位在汽车界有着传奇色彩的企业家，与有着"东方底特律"之称的汽车城重庆，会有怎样的深入合作？成为汽车行业内众说纷纭的话题。

在很长时间内，这个浙江企业家给公众的印象，与标准的中国传统制造密不可分。吉利不断围绕汽车制造扩展自己的版图，而最为人所熟知的就是完成收购沃尔沃的"蛇吞象"壮举。伴随着中国制造在全球的崛起，李书福也成为中国制造业的代表人物之一。

然而鲜为人知的是，在完成了一系列产业布局的同时，李书福与吉利汽车在智能制造方面的布局同样激进。

而下一个时代的李书福，是从重庆启程的。而且，李书福重仓重庆而来，早已不是什么新闻。

● 2021 智博会主展览厅广域铭岛展台
GYMD Digital Technology booth at the 2021 Smart China Expo main exhibition hall

在一年之内，吉利在重庆进行了几项重要布局：首先是吉利集团参与力帆汽车的重整，并在之后入主力帆，仅用半年，迅速为力帆摘星脱帽，并正式更名为"力帆科技"；其次是吉利集团的战略性高端新能源品牌极星，在重庆建厂投产；第三个大动作则通过 2021 智博会公之于众，那就是吉利集团的工业互联网全球总部广域铭岛落户重庆。

在 2021 智博会开幕首日的主展览厅，正对着入口的赫然便是广域铭岛"智能化柔性制造新模式"的数字孪生展台；在展台背后，是吉利汽车展示的掀翼式智能概念车；而在 2021 智博会开幕式主会场，李书福作为特邀嘉宾正在发表演讲。他在演讲中说："重庆是中国数字经济发展新高地，也是国家重要现代制造业的基地，工业发展历史非常悠久。这里不仅集聚了大量先进的制造企业，更有支持工业互联网、数字经济快速发展的政策与基础能力，是极具发展潜力的一片沃土。"[1]

显然，他这是把重庆当"家"了。

1 李辉，华龙网，《吉利控股集团董事长李书福：重庆是一片极具发展潜力的沃土》，2021 年 8 月 23 日。

第四节
回顾智博会这五年：时代赋予重庆的发展契机

我们很难为这个伟大的智能产业时代找到一个绝对的起点。因为无论是技术方面的累积，还是应用方面的推动，都是在循序渐进中产生的，犹如滴水穿石，绝非一日之功。

然而，当我们回顾智能产业的来时路，又总能发现一些特别的时刻。各种人物闪亮登场，各种技术喷薄而出，各种观点掷地有声，各种政策高屋建瓴。单看任何一个点，都仿佛是偶然的，但当一段时间内汇集了太多的偶然，恰恰宣示着一个全新时代隆重登场的必然。

2018—2022 年，对于人工智能来说就是无数个偶然汇集成必然的旅程，而在重庆市举办的智博会的发端与延续、萌芽与成长、升腾与沉淀、融合与收获，恰好也都发生在这段时间，与时代同频、与世界同行。

可以说，通过智博会的五年之旅，我们见证了世界从一个时代走向另一个时代，也见证了这期间重庆的发展。

一座城市的关键时刻

对区域经济稍有涉猎的人不难发现，不同的城市有着风格迥异的物产风俗，也有着截然不同的经济特征。为什么有的城市擅长工业制造，有的城市擅长农业养殖，有的城市服装供应全球，有的城市美食香飘全国？

一座城市屹立于全国乃至全球的城市海洋，其独特的产业掌纹，是由什么决定的？

在全球经济一体化到来之前，这个问题其实不难回答。靠山吃山，靠水吃水，这是人类祖先早已总结出的客观规律。但是当整个世界的经济产生了一体化连接，产业分工的范围从一条街不断扩大到一个村、一个镇、一个县、一个市、一个省、一个国家，最终扩大到全球范围，原本决定城市产业特征的气候优势、地理位置、物产资源等诸多条件，就逐渐失效了。

所谓"三分天注定，七分靠打拼"，主宰城市命运的因素，已经从靠天吃饭的自然条件转换为把握机遇的路径选择。

最初，在城市化进程中，往往依靠市场经济自发的实验形成一定的产业聚集。当然，这个过程充满偶然性，一个率先致富的个体户或

● 两江交汇形成的城市开放性格，使重庆乐于拥抱全球经济一体化

The open character of the city formed by the intersection of the two rivers makes Chongqing willing to embrace global economic integration

者一家偶然成功的企业，会给当地带来强烈的示范效应，从而引发产业集群。从纵向来看，成功的企业往往通过上下游配套，以市场化方式驱动产业链的完善；从横向来看，成功的企业往往会在本地产生样板效应，能启发越来越多的本土企业，从而形成更大的产业规模。

这种以点带线、以线带面的产业发展方式，往往是一座城市在城市化进程当中形成自己独有的经济特征的比较典型的逻辑。

然而在整个人类社会迎来革命性创新的重大历史关口，仅仅依靠城市自身纯粹自发的进化能力，去驱动一座城市快速地完成产业升级或转型，显然又是不切实际的。

怎样面向未来，把握趋势，展开想象，构建行动路线，去引导一座城市的转型，去改变一座城市的命运，这非常考验城市管理者的智慧。

通过高屋建瓴的规划，让一座城市依托于自身的产业基础，去拥抱趋势，完成城市产业的整体性提升，在改革开放四十年中成功案例无数。

20世纪90年代，上海抓住中国金融市场恢复的机遇，成为一座屹立全球的金融之城。

21世纪00年代，深圳抓住中国科技创新强国的机遇，成为一座影响世界的科技之城。

21世纪10年代，杭州抓住中国电子商务爆发的机遇，成为一座全球领先的电商之城。

面向21世纪的第三个十年，面对全球智能产业发展的新机遇，重庆的目标很清晰。

近100年来，重庆这座西部制造重镇，一直都在专注地扮演一个制造业基地的角色：从20世纪初的实业救国建厂萌芽，到抗战时期的工业西迁；从建国后的三线建设重工业沉淀，到20世纪80年代的轻工业崛起；从20世纪90年代的军工转民用、汽摩产业一枝独秀，

解码智能时代
从中国国际智能产业博览会瞭望全球智能产业（2018—2022）

到 21 世纪初汽车制造大爆发……

即便是在中国加入 WTO 之后，沿海城市的工厂如雨后春笋般出现，中国西部的制造业相对东部发展较慢，重庆也仍然没有失去制造重镇的地位。

而随着沿海制造业的产业调整，笔记本电脑、手机等精密电子制造开始在中国西部寻找新目的地的时候，重庆再一次抓住了机遇，承接了沿海制造业转移，引进了惠普、富士康、广达、英业达、旭硕、纬创和仁宝等一大批高端制造企业，完善了自身全类目制造业版图，为这座古老的制造业之城注入了现代基因。

所有的这一切，仿佛都在等待一个真正爆发的时刻到来，在那里，重庆将拥抱智能制造，站到一个面向未来的巨大产业"C 位"。

人工智能产业的发展，让这座城市发现，属于自己的更大的时代终于来了。

● 重庆这座古老的城市，正在注入越来越多的现代基因
　Chongqing, a traditional city, has injected more and more modern genes.

2022 年 7 月 4 日，智慧汽车问界 M7 正式发布，预售 2 小时订单破 2 万台，预售 4 小时订单破 4 万台，这是小康集团旗下赛力斯汽车与华为跨界造车深度合作的又一成果。

回看整个问界系列，合作双方一年前发布的 M5 战绩同样亮眼。自 2022 年 3 月 M5 启动交付以来，小康股份就迎来了爆发转折点，当月挺进高端新能源 SUV 销量榜前五名，交付第一个月就达到了 3000 辆，后续也一直维持新能源 SUV 销量排行榜前十名，87 天交付了 11296 辆，刷新了品牌单款车型交付破万最快纪录。[1]

小康与华为的合作，开创了智能汽车联合业务深度跨界融合的先河，进行了密切的新能源汽车领域合作，共同打造高性能、智能化移动出行解决方案，给用户提供更加高效便捷的智能汽车产品和智慧移动出行体验，把数字世界加载到每一辆车。

依托于重庆市在智能产业高屋建瓴的布局，以及智博会持续为本地制造企业与全球智能产业创造的链接机会，华为与小康的合作并非个例，越来越多的重庆制造正在迅速成为智能时代的创新先锋。在无数的企业智能化、数字化转型的同时，一个更为宏大的科学造城、造科学城的城市发展战略，正在进行。

2020 年 1 月 3 日，习近平总书记在中央财经委员会第六次会议上专题部署成渝地区双城经济圈建设，对成渝地区推进科技创新提出明确要求，指出要支持两地以"一城多园"模式合作共建西部科学城。[2]

随后，在重庆市高新区，西部（重庆）科学城的建设开启了高速模式：3 个月内，重庆市召开市规划委员会和市城市提升领导小组会议，会议审议了《中国西部（重庆）科学城国土空间规划（2020—2035 年）》；9 个月内，西部（重庆）科学城建设动员大会顺利召开，

1　严薇，上游新闻，《5 月产销数据陆续出炉 重庆车市焕发"新绿"》，2022 年 6 月 2 日。
2　新华网，《习近平主持召开中央财经委员会第六次会议》，2020 年 1 月 3 日。

集中开工科学大道、科学谷、科学城生态水系示范工程等 79 个重点项目，标志着西部（重庆）科学城全面启动建设；10 个月内，重庆高新区举行西部（重庆）科学城高校、科研院所重点创新平台项目集中签约活动，此次签约活动共签约 24 个项目……

自西部（重庆）科学城启动建设以来，立足实现高水平科技自立自强，一大批大装置、大院所、大平台、大项目持续加速推进：中国自然人群生物资源库招募样本采集人群超 10 万人；北京大学重庆大数据研究院 1 个数字化转型促进中心、13 个大数据智能化相关前沿实验室正加速科研攻关；中科院汽车软件创新平台落地建设、核心科研团队已经入驻；上海交大重庆人工智能研究院正式落地……[1]

21 世纪的第三个十年，西部（重庆）科学城正在抓住一切机会，加速建设具有全国影响力的科技创新中心核心区。

狭隘地看，人工智能是技术性的，无论是语音识别、视觉识别，还是深度计算、神经计算、边缘计算、大数据，都需要拥有强大的技术基因。

但从更为宏观的视角来观察，能否在智能时代形成具有竞争力的智能产业聚集，更为核心的是产业基础。毕竟技术可以迁移和流动，而产业则更需要聚合与扎根。

从技术到产业的应用路线来讲，先进技术投入到生产制造的改善，远比融入最终的产品速度更快、效率更高。

因此对于人工智能与制造业的结合，重庆是绝佳的目的地和试验场。

---

1 刘翰书，上游新闻，《加速建设具有全国影响力的科技创新中心核心区 西部（重庆）科学城金凤实验室本月投用》，2022 年 5 月 12 日。

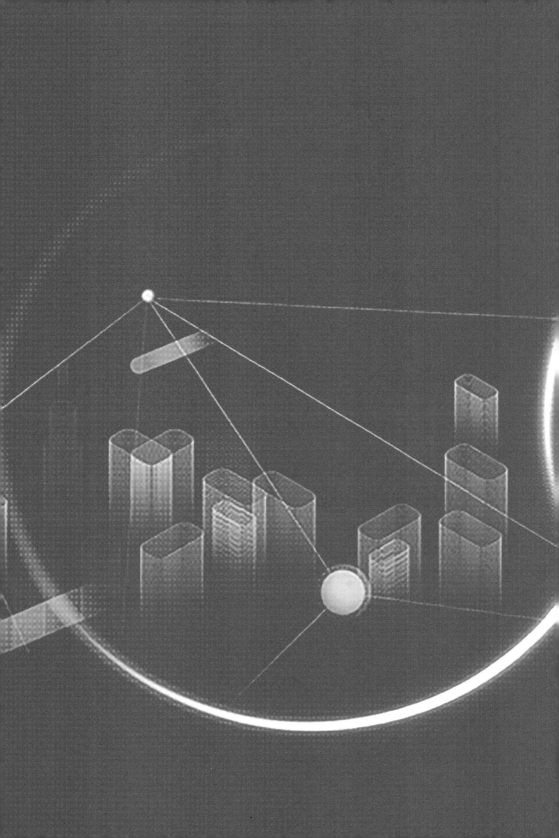

第二章

智能经济：万千产业问题的一个标准答案

　　每一种产业发展到极致，都会遇到相似的问题；每一个时代行进到终点，都会遇到相似的瓶颈；而每当经济社会积累了足够丰富的问题、遭遇了足够顽固的瓶颈，现有技术的迭代已臻极致，过往经验的喧嚣突然静寂，恰在此时，也总有新一次工业革命的横空出世。

　　规模化生产与个性化定制之间的矛盾，物理世界与虚拟世界之间的裂痕，实体经济与数字经济之间的缝隙，数字意识与社会算法之间的差值，经济社会的万千产业问题，终于成功向时代索取到了一个新的标准答案——智能化经济。

第一节
解决产业问题，全球发展智能产业的共识

2018 年 8 月 23 日至 25 日，重庆国际博览中心，首届中国国际智能产业博览会正在如火如荼召开，一直关注自动驾驶技术发展的 65 岁重庆南岸区市民陈文国，终于在家门口得偿所愿 —— 看到了真实的自动驾驶汽车。

除了自动驾驶汽车，可以刷脸支付的自动售货机、VR 技术模拟潜水参观水下博物馆、3D 打印、无人机飞行……同样令陈文国十分惊叹。和他一同流连智博会现场，惊叹于智能时代"黑科技"的观众，据统计超过了 50 万人次[1]，甚至为了响应大家的热情，组委会做出了延展一天的决定。

如果穿越时空，会发现相似的场景也在距离重庆近 8000 公里的德国汉诺威上演过。

时间倒回到 70 多年前，1947 年 8 月 18 日至 9 月 7 日，首届汉诺威工业博览会在德国汉诺威举行，人们被当时展会上世界最小的柴油引擎、假牙、可折叠婴儿推车和甲壳虫小轿车等吸引，并想象他们将会为人类生活带来的重大影响。

汉诺威工业博览会自 1947 年起，每年一届，已成为国际技术和工业的交流平台，就像中国国际智能产业博览会象征着智能产业时代

1　杨野，上游新闻，《50 万 +！这数字让阿里、腾讯都惊呆了》，2018 年 8 月 26 日。

正加速到来，汉诺威工业博览会也有其标注时代的意义——它是电气化工业时代、自动化工业时代的产物，也正是在 2013 年的汉诺威工业博览会上，"工业 4.0"即第四次工业革命概念被正式推出，智能产业时代的帷幕正式拉开。

不同表述下的产业共识

"工业 4.0"概念萌生于 2011 年，概念提出者是德国国家科学与工程院前院长亨宁·卡格曼和德国人工智能研究中心前主任沃尔夫冈·瓦尔斯特。

如何理解"工业 4.0"？

● 德国最先提出的"工业 4.0"概念，后来也被称为"第四次工业革命"

The concept of "Industry 4.0" first put forward by Germany was later called "the Fourth Industrial Revolution".

概念提出者之一的亨宁·卡格曼认为："需要从两个方面来理解，一方面是万物互联，即互联网与生活中的很多环节都可以发生关联，另一方面是物理世界与虚拟世界的融合，即虚拟层面上的改变可以立即影响到物理层面，反之亦然。"[1]

"工业 4.0"有更贴近生产、更为具体的诠释，莱比锡商学院院长平克瓦特认为："联网化、数字化、智能化等，是实现'工业 4.0'的手段而非目标，根本的目标其实是从根本上改变现有的生产思维模式，从大批量生产转向以客户需求为导向的小批量定制化生产，高效率、高质量地完成以人为本的个性化消费，并控制成本。'工业 4.0'的任务，不是单纯的工业生产的网络数字化升级，而是将产品、消费与生产三方融入到一个可以互相沟通的网络中。"[2]

上述内容都是专家对"工业 4.0"的早期解读，解读的角度不同，释义也就不同，且随着时间的推移、新技术的发展演变，实际上"工业 4.0"的内涵也在不断变化和丰富。但有一点不会变，这点体现在德国"工业 4.0"工作小组于 2013 年向德国联邦政府提交的报告的名称中。报告《保障德国制造业的未来：关于实施"工业 4.0"战略的建议》，正如其名称中指出的，"工业 4.0"其实代表了"制造业的未来"，那些擘画制造业未来的生产资料、生产方式、技术革新、经营理念等，都在"工业 4.0"的诠释范围。

"制造业的未来"在德国被表述为"工业 4.0"，而在其他国家，则有不同的表述。美国称之为"工业互联网"，日本称之为"互联工业"，中国则主要以"智能制造"居多。这是不同国家对各自制造业未来的表述，其实各国对大数据、人工智能、5G 等新技术、新基建的需求是相同的，但基于不同的经济体量、工业基础、产业理念等，又有各自的侧重点。

1 冯雪珺、管克江，人民日报，《汉诺威工博会呈现"数字化世界"》，2015 年 4 月 15 日。
2 冯雪珺、管克江，人民日报，《汉诺威工博会呈现"数字化世界"》，2015 年 4 月 15 日。

● 重庆机电集团展出机电装备智能化生产新模式
Chongqing Machinery & Electronics Holding (Group) Co. Ltd. exhibits a new model of intelligent production of electromechanical equipment.

有个共识是，德国的"工业 4.0"侧重于设备的智能提升和智能化标准方面，这与德国工业"强设备"及设备输出的工业原有路径相符。而中国的智能制造，强调信息技术与制造业的深度融合，相比而言，中国在产业链的视角更广，不仅涉及生产端，也重视消费端与生产端的信息化融合，落脚点在提升制造业的整体水平，在于产品生产的质量、效率、成本层面，背后则是中国属于制造业大国的基本现状。

无论表述如何，无论侧重点如何，道路和沿途的风景或有差异，但基于新的技术与理念，发展智能产业，已成全球主要国家的共识。

智能产业的驱动力

在提出"工业 4.0"的十年后，智能产业已今非昔比，进入蓬勃发展的快车道。

有个疑问是，发展智能产业为什么会成为全球范围内的共识？它的驱动力是什么？它为何如历史潮流，浩荡不可阻拦？

这里面的原因当然是复杂的，也是分层次的，有浅显的市场需要，也有深层的竞争驱动，有客观条件成熟的自然萌发，也有主观意识思考的未来渴望。

对智能产业的共识，首先是对制造业的共识。2008 年国际金融危机虽然已经过去 10 多年，其影响还在发挥作用，而且不限于金融领域，对制造业行业的影响更为深远。有观点认为，制造业空心化、虚拟经济占据主要地位不利于经济稳定，也是高失业率的罪魁祸首。

一年后的 2009 年 11 月，美国前总统奥巴马提出，美国经济要从过去金融信贷上的高消费模式转型，要重新平衡制造业和服务业——"脱虚向实""虚实并进"的经济政策思潮，让制造业再一次回到聚光灯下，也持续主导着后续的全球经济竞争格局。

靠制造业抵抗经济风险是一方面，这是社会经济的压舱石；另一方面，发展智能产业，自身能创造巨量经济产值。中商产业研究院预测，中国智能制造装备市场规模 2022 年将超 2.6 万亿元[1]，这个数据在 2019 年时不到 1.8 万亿元[2]，可谓增长迅速。

德国"工业 4.0"可以对德国经济产值产生多大影响？预计 2025 年，可使德国在工业制造、汽车、化学、IT、电子、农业 6 个行业的产值增加 787 亿欧元，使德国整体产值增加 2670 亿欧元。[3] 这个产值增加值，孤立地看并不惊人，但若与 2021 年德国 GDP 35639 亿欧元的总量对比，不难发现这给德国经济带来了很大比重的增量。

除了对经济的促进，推动智能产业发展的主观驱动力，还包含国

解码智能时代
从中国国际智能产业博览会瞭望全球智能产业（2018—2022）

1　经济观察报，《传统质检人眼变"智能眼"联想算法赋能企业转型再获奖》，2021 年 12 月 29 日。

2　青岛财经日报，《我国开启新一轮智能制造施工大幕》，2021 年 04 月 23 日。

3　王罗汉；王伟楠，《全球科技经济瞭望》第 36 卷 2021 年第 12 期，《德国工业 4.0 十年发展回顾，中国能借鉴什么？》。

与国之间的竞争与合作。

早在 2014 年，国际科学界就对"工业 4.0"进行了概念的解读："'工业 4.0'这一名称的含义是人类历史上的第四次工业革命。"[1] 而同样在 2014 年，中德双方签署的《中德合作行动纲要》，就对"工业 4.0"与"智能化时代"之间的关系，进行了清晰的定义："工业 1.0 是蒸汽机时代，工业 2.0 是电气化时代，工业 3.0 是信息化时代，工业 4.0 则是利用信息化技术促进产业变革的时代，也就是智能化时代。"[2] 而所谓第四次工业革命，是以人工智能、清洁能源、自动控制技术、量子信息技术、虚拟现实为主的全新技术革命，就像蒸汽时代之于英国，电气时代之于德国，信息技术时代之于美国，其携带的能量必然引起世界竞争和合作格局的转变，没有谁愿意错过这一轮的工业浪潮，也没有谁愿意在全球化的当下失去产业链上的地位。

如果将目光再聚焦一点，智能 + 制造构成的智能产业，能解决不少现实而具体的问题。

在"工业 4.0"概念之父亨宁·卡格曼的一段表述中，这些问题有集中体现："工业 4.0"的设想是，生产具有高度灵活性，产品个性化、节约资源，年长的工人能得到智能辅助系统的帮助；生产人性化，工人可在家或就近上班。智能工厂并非无人工厂，它只是拓宽了工人参与和决定的空间。直接在车间一线工作的人减少，但其他人可在系统方案、研发、协调方面参与工作。

"生产灵活性，产品个性化"，能够解决产业中的产能过剩问题。罔顾市场真实需求情况和需求量进行无节制的生产，是导致产能过剩的主要原因。智能化的生产方式，实现了人与设备、设备与设备的互联，还将数据收集、共享的触角拓展到需求端，从而实现更为精细化的生产。

1　高野敦，人民网，《【名词解释】工业 4.0：德国欲掀起第四次工业革命》，2014 年 1 月 13 日。
2　李松涛，中国青年报，《李克强访欧提"工业 4.0"，对中国正当其时》，2014 年 10 月 14 日。

智能产业设备自动化程度高，设备自我决策能力强，将在生产环节减少对人的依赖，改变工人参与制造的时间和空间。一方面，能够应对劳动力成本上升的问题，降低成本；另一方面，也能够应对全球各国就业人口减少的趋势。

同时不能忽视的是，每一次产业变革，都与能源息息相关。能源改变生产方式，生产方式反过来也谋求能源类型的变革。蒸汽时代、电气时代，能源的主要支撑是煤、石油这类不可再生的资源，且对环境并不友好，如今这个支撑在松动，但依旧是整个人类社会运转的主要动力来源。

智能产业如何解决能源问题？一个关键词是节约，另一个关键词是替代。当生产更为精准，互联互通更为畅快，自然更加节约能源。

● 参展观众在长安汽车展位现场体验智能数字座舱
People experience intelligent digital cockpit in Chang'an Automobile booth.

而对能源的替代，我们可以看到光伏、新能源汽车等产业，将减少智能时代人类对不可再生能源的依赖。

罗马不是一天建成的，智能产业的大门也不是在某一年被猛然推开的，产业的发展有时候并不完全遵循人类的意志，它有一种自发性，尤其当充分具备技术要素的时候。机器人、互联网、物联网、大数据、云计算等大量新技术的发展和成熟，是打开智能时代大门的力量，也是构筑智能时代的拼图。

智能产业的发展，叠加着层次丰富的主客观因素，带着一种宿命感，如历史潮流，浩荡不可阻拦。

第二节
布局智能战略，中国进行时代红利的切换

如果说"工业 4.0"的提出，第一次从概念上引出了智能时代，那么智能时代的实质性萌芽发生在什么时候？

2008 年可能是一个比较容易达成共识的答案。

2008 年注定是不平凡的一年，这一年发生了不少重大的事件，比如全球经济危机和北京奥运会。其实这一年，有三个"超越"指向一个时代的开启。

在 2021 智博会的开幕式主旨演讲中，中国电子科技集团有限公司总经理、中国工程院院士吴曼青，提到 2008 年世界上同时发生了三次超越："一是城市人口超过了农村人口，城市人口超越意味着人的集结；二是机器链接数量超过人类链接的数量；三是移动链接的数量超过了固定链接的数量。"

鲜活的数字世界，充满想象的智能时代，中国的布局是什么？如果说布局是主观愿望，那么发展智能产业的客观条件又是什么？优越的客观条件其实就是我们常说的"红利"，发展智能产业，中国的"时代红利"又是什么？

智能产业战略布局

中国、美国、德国在世界经济和制造产业中占有举足轻重的地位，既有发展智能产业的共识，也在智能产业战略的布局与实践上最为前沿、最为完备。

每个国家的智能产业战略一定是基于本国的实际情况提出的，不同的战略路线，既互为补充，又互为参照。通过梳理美国、德国的智能产业战略布局，或许能够更为清晰地理解中国的智能产业战略。

美国的智能产业布局始于对制造业的回归。2009 年 12 月、2010 年 8 月，奥巴马相继签署了《美国制造业再振兴法案》和《制造业促进法案》。两个法案的关键词是制造业，揭示了一个显而易见的道理——智能产业是制造业 + 智能，如果制造业空心化，智能产业就会沦为无根之萍。

随后美国开始布局"智能制造"，比如以法案形式确立《国家制造业创新网络宪章》，旨在主张建立关键领域的研究所来聚合产业界、学术界、联邦及地方政府等多个主体，建立和完善创新生态系统，而《美国先进制造领先战略》则旨在连接中小企业参与智能制造，并创建生态系统。两个方案的关键词都是"创新"，前一个是激发创新，后一个是将创新落地并赋能广大中小企业。

2021 年 6 月，美国参议院通过了《2021 年美国创新和竞争法案》，主张美国联邦政府应通过关键领域的公共投资增强美国新技术实力，芯片、5G 网络、量子计算和信息系统等都属于关键领域范畴。这里的布局关键字是"关键领域"投资，也就是智能产业发展要重点突破。

德国最早提出"工业 4.0"概念。2013 年，德国正式发布了《保障德国制造业的未来——关于实施工业 4.0 战略的建议》，将"工业4.0"上升到国家战略层面。

德国布局智能产业频频提及的关键词是"数字化"。2016 年，德国发布《数字化战略 2025》，发展数字化技术、可信赖的云、德国数据服务平台、中小企业数字化、进入数字化等 12 项内容。在 2019 年 11 月发布的《德国工业战略 2030》中，除了改善工业基地的框架条件、加强新技术研发和调动私人资本等内容，德国认为当前最重要的突破性创新依旧是加快数字化进程。

中国的智能产业战略布局可追溯到 2013 年中国工程院、国家工信部等启动的"制造强国战略研究"重大咨询项目及其后提出的十个"重点领域"，勾勒出我国智能产业布局的关键，包括：新一代信息技术产业、高档数控机床和机器人、航空航天装备、海洋工程装备及高技术船舶、先进轨道交通装备、节能与新能源汽车、电力装备、农机装备、新材料、生物医药及高性能医疗器械。

2021 年 12 月 21 日，工信部联合国家发展改革委、教育部、科技部等部门发布了《"十四五"智能制造发展规划》，提出了三项 2025 年的主要目标。

三项主要目标包括：①转型升级成效显著。70% 的规模以上制造业企业基本实现数字化网络化，建成 500 个以上引领行业发展的智能制造示范工厂。制造业企业生产效率、产品良品率、能源资源利用率等显著提升，智能制造能力成熟度水平明显提升。②供给能力明显增强。智能制造装备和工业软件技术水平和市场竞争力显著提升，市场满足率分别超过 70% 和 50%，培育 150 家以上专业水平高、服务能力强的智能制造系统解决方案供应商。③基础支撑更加坚实。建设一批智能制造创新载体和公共服务平台。构建适应智能制造发展的标准体系和网络基础设施，完成 200 项以上国家、行业标准的制订修订，建成 120 个以上具有行业和区域影响力的工业互联网平台。

同时，围绕创新、应用、供给和支撑等四个方面，《"十四五"智能制造发展规划》部署了智能制造技术攻关行动、智能制造示范工厂

● 重庆四联测控技术有限公司展出仪器仪表行业 5G+ 工业互联网场景
Chongqing Silian Measure & Control Technology Co., Ltd. exhibits the scenario of 5G instrument industry+Industrial Internet.

建设行动、行业智能化改造升级行动、智能制造装备创新发展行动、工业软件突破提升行动、智能制造标准领航行动等六个专项行动。

《"十四五"智能制造发展规划》以发展先进智能制造业为核心目标，规划制造强国的推进路径。基于国家对智能制造的大力支持，我国智能制造行业保持着较快的增长速度。与此同时，我国智能制造产业已从初期的理念普及、试点应用进入多级推进、协同创新的新阶段。

智能产业的时代红利

中国文字中的"势"字，与"红利"一词在含义上有十分巧妙的联系。

"势"的古字，意象为圆球处于高土墩的斜面即将滚落。如果将智能产业视为圆球，红利则是斜面，斜面越长、落差越大，圆球最终能产生的能量也越大。

构成中国智能产业"斜面"的红利是什么？为什么说中国发展智能产业是"顺势而为"？

前一个制造业时代，中国的劳动力数量和劳动力成本曾是一个巨大的时代红利。进入智能时代后，中国的劳动力红利其实并未消失，只是说法有些变化，叫人才红利，也就是工程师红利。

中国有数量庞大的敬业的工程师群体，长江商学院甘洁教授曾说："在慕尼黑找物联网相关的软硬件工程师，可能招聘广告发出来半年都找不到人。如果在美国，要么花高薪去挖例如苹果大公司的人才，或者去找五六十岁的老工程师。西门子曾出过内部报告，中国工程师每年工作时间是外国工程师的两倍。"

世界工程组织联合会主席龚克也曾说："中国每年工学类普通本科毕业生超过 140 万人，中国工程师联合体目前共有 76 家发起成员单位，将团结中国 4200 多万工程科技人才，工程师红利已成为推动中国经济高质量发展的重要力量，也让中国有能力成为世界工程发展的重要驱动力。"[1]2001 年中国加入 WTO，海量低成本的劳动力是推动中国成为世界工厂的关键比较优势，这在广泛意义上被称为中国的"人口红利 1.0"。二十年后，当中国工厂车间里的劳动力不断减少，有人开始质疑中国制造的后续生命力时，工业制造又迎来了从依赖工人转而依靠自动化设备的智能制造时代。而曾经在车间里挥汗如雨的工人们，他们的子女得到了良好的教育，成为这个国家工程科技领域的"人才红利 2.0"。

从人工制造到智能制造，从人口红利到人才红利，中国制造的自

---

1 李鲲、宋瑞，瞭望，《瞭望｜龚克：工程师红利亟需重新认识》，2021 年 5 月 31 日。

● 中国的"人口红利"，正在智能时代转换为全新的"人才红利"

China's "demographic dividend" has been transformed into a new "talent dividend" in the age of intelligence.

我进化能力让全球叹为观止。

除了人才红利，还可以从上文提及的美国、德国的智能产业战略布局的关键词中寻找中国智能产业的时代红利。

"中国制造体量大，2012 年到 2020 年，中国工业增加值由 20.9 万亿元增长到 31.3 万亿元，其中制造业增加值由 16.98 万亿元增长到 26.6 万亿元，占全球比重由 22.5% 提高到近 30%。中国工业拥有 41 个大类、207 个中类、666 个小类，是世界上工业体系最为健全的国家，在 500 种主要工业产品中，有 40% 以上产品的产量世界第一"[1]，这是中国制造基本面的产业红利，在全球占据着重要的位置。

"数据化"是指大数据、人工智能、移动互联网、云计算等一系

1　周頔、孙铭蔚，澎湃新闻，《工信部：我国制造业增加值连续 11 年位居世界第一》，2021 年 9 月 13 日。

列由数字化技术组成的"数字综合体"。中国拥有全球最多的人口和全球最大的市场，互联网的普及使得中国成为全球数据量最大的国家，因而在智能产业发展所需数据量方面有相当大的优势。这是智能时代到来之际最为关键的数据红利。

同时，正如亨宁·卡格曼所指出的那样，"数字化基础设施建设是下一场工业革命的关键，建设网络和扩大带宽十分重要"[1]。截至2021年6月，全国建有5G基站96.1万个，5G网络覆盖全国所有的地级市、95%以上的县域地区、35%的乡镇地区，这算是智能产业发展在数字化方面的基建红利。

从人口红利到人才红利，这是红利的进阶，让上一代制造业中的人口红利带动制造业整体发展和地位提升，为新的智能产业奠定基础，这是用红利换红利。

中国智能产业发展可享受的红利很多，短板也不少，好在整体趋势就像圆球从"斜面"出发，势能越来越大，势头越来越好。

如果将眼界拉远放宽，以更高更广的角度审视智能产业本身，智能产业本就是时代红利，谁重视智能产业、谁发展智能产业，谁就能在发展上享有更多的机会和可能。

解码智能时代
从中国国际智能产业博览会瞭望全球智能产业（2018—2022）

---

1 胡小兵，新华网，《"当机器与网络连接将发生下一场工业革命"》，2014年04月09日。

第三节
打造"智造重镇"，重庆完成竞争优势的升级

滚滚长江，悠悠嘉陵，重庆的工业化之旅经历了上百年的时光。

抗日战争时期沿海工业内迁，建国后三线建设时期核心工业内迁，以及全球化大背景下的沿海产业结构调整推动电子工业内迁，近百年来，中国三次规模较大的制造工业迁徙，重庆都抓住了机遇。

2017 年底，重庆提出以大数据智能化引领产业转型升级，并确定了大数据、人工智能、集成电路等 12 个智能产业重点发展领域，随后 2018 智博会在重庆召开并永久落户。重庆的制造业从此注入了智能因子，插上了智能翅膀，开始了新的腾飞。

重庆的制造业蜕变之旅，已经重新启程，加快建设"智造重镇"，完成重庆竞争优势的升级，注定更为波澜壮阔。

数字产业化

在 2021 智博会开幕式上，重庆再一次传递出清晰的信号：坚持一手抓数字产业化、一手抓产业数字化，积极推动数字经济和实体经济深度融合，全面加强战略谋划部署，加快建设新型基础设施，倾力打造智造重镇，精心培育智慧名城。

数字产业化的释义是：数字技术带来的产品和服务。如电子信息

制造业、信息通信业、软件服务业、互联网业等，有了数字技术后才出现的产业。

数字产业化方面，着力构建"芯屏器核网"全产业链，背后承载着五个庞大的产业集群：集成电路产业集群、新型显示产业集群、智能产业集群、核心器件产业集群、工业互联网及软件产业等信息服务业集群。

"芯屏器核网"在重庆纷纷落地，遍地开花。

芯，重庆是国内最早发展大规模集成电路的城市之一，已初步建成"IC 设计—晶圆制造—封装测试及原材料配套"全流程体系，已有集成电路重点企业 63 家，设计类企业近 40 家。聚焦到企业，华润微电子建设的 12 英寸晶圆生产线，正带领企业实现制造能力的突破，联合微电子中心面向全球，发布了 130nm 成套硅光工艺 PDK 等三套工艺 PDK……

屏，重庆近年来初步建成了"玻璃基板—液晶面板—显示模组—整机"的全产业生态圈，形成了"一核三园"的发展格局，产值规模居全国前列。主要企业有京东方、惠科、莱宝、康宁等，包括京东方第 8.5 代液晶面板、惠科金渝第 8.6 代液晶面板两条重要生产线以及京东方第 6 代 AMOLED（柔性）面板和康佳半导体光电产业园两个重点项目。

器，重庆智能终端产业形成了"品牌多元、代工多家、配套多样、产品多类"的产业体系。笔记本电脑产业已汇聚广达、英业达、仁宝、和硕、纬创、富士康等全球六大代工厂，连续 7 年稳居全球最大笔记本电脑生产基地；手机产业已落户 OPPO、VIVO、传音等 3 家全球出货量排名前 6 位的手机品牌商，产量约占全球的十分之一；入驻了京东方、莱宝、联创电子、惠科等智能终端零部件企业 1000 多家，规模以上企业达 220 余家。年产 PC 约占全球 30%，手机约占全球 10%，智能手表占全球 20% 以上。新型终端落户了生产智能投

解码智能时代
从中国国际智能产业博览会瞭望全球智能产业（2018—2022）

● 2021 智博会上，峰米科技发布新产品
Formovie released new products at the 2021 Smart China Expo.

影终端产品的小米生态链企业蜂米科技、专注家庭地面和立面的清洁机器人的福玛特……

核，在智能传感器领域，重庆现有企业约 60 家，传感器产业规模近 200 亿元，初步形成"材料＋设计＋制造＋封测＋集成"的传感器产业生态链，在温度、压力、流量等基础工业传感器领域实现了全覆盖。2020 年，西部（重庆）科学城北碚园区获全市唯一"重庆市传感器特色产业基地"授牌，在 2020 世界半导体大会·全球传感器与物联网产业创新峰会上发布的《2020 赛迪传感器十大园区》白皮书中，重庆（北碚区）传感器特色产业基地名列全国第九位，也是西部地区唯一入选的产业园。同时，重庆致力于完善新能源和智能网联汽车产业生态，发展"大小三电"、智能控制系统等核心零部件，以两江新区、永川区为核心区域，形成了以长安体系为龙头，10 多家整车企业为骨干，上千家配套企业为支撑的"1+10+1000"的优势汽车产业集群。

网，重庆工业互联网发展态势良好。作为工业互联网标识解析国家五大顶级节点之一，重庆已上线并接入二级节点 19 个，标识注册量已经破 41 亿，标识日解析量超 1400 万，数据提升速度位列全国第一，累计建成 4.9 万个 5G 基站，形成覆盖主要工业制造区域的高效、稳定网络体系，成为全国首批 5G 规模组网试点城市。工信部评选的 15 家跨行业跨领域工业互联网平台中，有 11 家在重庆布局，重庆已集聚平台服务、解决方案、大数据服务等企业近 200 家，累计推动近 10 万户企业"上云上平台"；累计实施逾 3000 个智能化改造项目……具体项目方面，扎根重庆的中冶赛迪，推出钢铁行业首个工业互联网平台，中国工业互联网研究院重庆分院、国家工业大数据中心重庆分中心打造的汽摩配、电子制造工业互联网公共服务平台已投用……

对于《重庆市科技创新"十四五"规划》明确提出要面向"智造重镇""智慧名城"建设，一手抓研发创新，一手抓补链成群，着力构建"芯屏器核网"全产业链，中国工程院院长李晓红曾表示："我们中国工程院约三十多位院士先后两轮论证，对这个规划给予了高度评价，认为是未来重庆市科技创新发展的一个风向标，甚至是科技创新的引擎。"

作为重庆智能产业里的 5 大核心，近年芯屏器核网全产业链的快速发展，构筑起重庆新的发展格局和竞争优势。

产业数字化

作为重要的制造业基地，重庆有深厚的产业基础。在国内 41 个工业门类中，重庆拥有 39 个，可谓工业门类齐全。重庆制造业数字化转型具备良好的产业基础，而"产业数字化"也为重庆带来了新的

解码智能时代 从中国国际智能产业博览会瞭望全球智能产业（2018—2022）

战略机遇。

产业数字化的魔法已经在重庆的汽车和摩托车制造行业生效。

"在重庆长安汽车两江工厂，通过制造数据的全过程赋能，构建了敏捷、智慧的多车型柔性生产线，下线一辆高品质汽车仅需 52 秒；在金康赛力斯两江智慧工厂，支撑客户个性化定制整车颜色的生产喷涂系统，切换时间仅需 15 秒。"[1]

摩托车生产企业宗申集团推出了忽米网工业服务平台，并率先将其布局在宗申摩托车总装 1011 生产线上。

"在这条生产线上，机器手臂来回穿梭，传感器实时对产品进行检测，无需人工操作；从上线到包装所有环节均实现数据自动采集，后台据此可自主编排生产计划……该生产线用工数量减少了一半，人均产出却提升了 2.2 倍，自动纠错防错能力还大幅提升了 10.6 倍，实现了平均 10 秒就下线一台发动机。"[2]

这样的"产业数字化"改造，正大量地在重庆开展着。来自市经济信息委的数据显示，2021 年，重庆推动实施智能化改造项目 1295 个（累计 4075 个，直接带动工业投资 600 多亿元），认定智能工厂 38 个（累计 105 个）、数字化车间 215 个（累计 574 个）。[3]

另外，经测算，全市 1500 余家实施智能化改造的规上企业产值平均增长 46.8%，对全市工业产值贡献超 60%；数字化车间、智能工厂示范项目生产效率平均提升 59.8%。

数字经济对重庆区域经济的贡献也越来越大。据统计，2018 年至 2021 年，重庆市数字经济分别增长 13.7%、15.9%、18.3%、15%，常年维持高增速，2021 年，重庆数字经济增加值突破 7300 亿元，占

---

1 人民日报，《2021 智博会：数字经济助力重庆高质量发展》，2021 年 08 月 21 日。

2 栗建昌、何宗渝，新华每日电讯，《"智慧公园"浓缩重庆"智造"，智能产业 3 年奔万亿》，2019 年 12 月 23 日。

3 梁浩楠，华龙网，《聚焦智能化、数字化、绿色化 2022 年重庆将加速"四换步伐"提振工业有效投资》，2022 年 4 月 27 日。

地区生产总值的比重提升至 27.2%，跻身全国数字经济第一方阵。

在 2021 智博会上，国家工业信息安全发展研究中心发布了《重庆市两化融合发展数据地图（2021）》，显示重庆市两化融合发展水平连续多年位居中西部第一，高新技术产业和战略性新兴产业对工业增长的贡献率分别达 37.9%、55.7%。

报告中，引人注意的还包括：14.2% 的重庆制造企业初步具备探索智能制造基础，汽车摩托车、电子信息、装配、材料化工、医药五大行业融合发展成效显现。

这说明，重庆的关键性行业、引领性制造企业，已融入智能时代的潮头，实现了竞争优势的升级，正要带领重庆智能产业扬帆起航、逐浪前行。

第四节
过去的五年，成为角逐未来数十年的关键

　　如果时间倒回 2019 年 8 月 26 日至 2019 年 8 月 28 日，在关于 2019 智博会的诸多报道中，会发现"第一方阵"四个字出现的频率非常高。

　　这四个字出自中共中央政治局委员、国务院副总理刘鹤在展会开幕式发表的重要讲话，他指出：在各方积极因素推动下，重庆经济社会发展取得了显著成就，重庆智能产业的发展已经位于全国第一方阵。[1]

　　在重庆智能产业史上，2019 年多少有些承前启后的意味：一来这是中国国际智能产业博览会永久落户重庆的第二年，展会与重庆智能产业的化合反应正在加剧；二来，正如上文所说，这一年重庆智能产业发展进入全国第一方阵，到了发展条件越来越完备、发展势头越来越强劲的时候。

　　可以说，自 2018 年重庆举办智博会始，至 2019 年重庆智能产业发展进入快车道，再到 2022 年，这 5 年时间里重庆把握住高水平建设"智造重镇""智慧名城"的历史机遇，为未来数十年的发展指明了方向、夯实了基础。

1　韩政、陈翔，重庆晨报，《重庆智能产业发展已处"第一方阵" 下一步目标建智造重镇、智慧名城》，2019 年 8 月 30 日。

智博会的"智"与"实"

永远不要低估展会对城市甚至国家的影响，也不可小视展会与城市或国家相关产业相辅相成的关系。

德国走向世界制造业强国，汉诺威工业博览会功不可没。每年在拉斯维加斯举办的国际消费类电子产品展览会，既扩大了美国在消费类电子产品及技术领域的全球影响力，展会上的全球知识、人才、产品和技术，也推动着美国在该领域的纵深发展。

2022年，是智博会落户重庆的第五年。5年时间，总结智博会对中国智能产业的影响，或许时间尺度还太短；但5年时间，足以展现智博会为重庆这个城市带来的变化和为未来发展奠定的关键基础。

主要分"智"与"实"两个方面。

"智"，按照它本身的含义，即专家学者的观念、实践者的技术路线和流派，甚至概念产品、尖端产品的呈现，都为重庆及智能行业带来了"智"。这个"智"，是推动重庆未来几十年智能产业发展直接的"智"。

还有间接的"智"，引申过来，是"道路"和"共识"。什么道路呢？大力发展智能产业这条道路；什么共识呢？大力发展智能产业这个共识。

某种程度上，"道路"和"共识"至关重要。

"道路"指明方向，事关一个城市的发展战略，是城市发展的顶层设计；"共识"则决定能够为这条道路聚集多大的力量。经过智博会5年的宣传和教育，重庆已经在市民及全世界关注智能产业的人们心中深深打上了智能制造的烙印，这便是共识。这个共识将吸引本地及世界各地的产业相关者，在重庆进行相关研究、投资、创业等，从而推动产业迅猛发展。

这是五年来智博会为重庆未来发展带来的"智"的部分，还有

● 2021 智博会 "5G+ 智能制造" 无人化供应链共享协同平台项目合作签约仪式
Smart China Expo 2021 "5G+Intelligent Manufacturing" Unmanned Supply Chain Sharing Collaborative Platform Project Cooperation Signing Ceremony

"实"的部分。

2018 年，智博会官方数据显示：按照现场集中签约和场外专场签约相结合的方式，共签约重大项目 501 个。其中，现场集中签约 56 个项目，包括来自重庆 20 个区县、开发区的 36 个市内签约项目；来自新加坡、湖北省等国家和地区的 10 个国际和兄弟省市签约项目；以及重庆市政府、有关部门与国内外知名企业、机构现场签约的 10 个战略合作项目。此外，重庆各区县、开发区，和兄弟省市进行场外专场签约 445 个项目，覆盖国内外大数据智能化创新各行业领域[1]。

从 2018 年的签约数据可以看出，重庆各区县展开了与国际、各省市、各企业、各机构在智能产业方面的全面合作。虽然这些签约项

1　曲鸿瑞、刘翰书，上游新闻，《首届智博会签约重大项目约 501 个，合计投资约 6120 亿元》，2018 年 8 月 23 日。

目或许并非仅因智博会促成，但不可忽视智博会在其中的平台作用、虹吸作用、激发作用。

2019年至2021年，根据四届智博会公开的项目签约情况统计，重庆共计签约相关重大项目1194个。

可以想象，随着聚焦新一代信息技术、智能制造、智能服务三大板块，覆盖集成电路、物联网、大数据、新能源和智能汽车、智能硬件、智能装备、智能工厂、智慧总部、智慧物流等众多智能产业和智能化应用关键领域的1300多个项目，近2万亿元的投资金额，逐渐在渝州大地落地生根，释放效用，构筑起智能产业的庞大基础并发挥聚集作用，将重庆打造为国内乃至世界的智能产业新星。

这是智博会为重庆带来的"智"与"实"。当然，过去5年，并非只有智博会为重庆智能产业发展做标注、做引领，实际上，未来发展的关键，还藏在政策的顶层设计、产业链的规划、创新氛围的营造上。

角逐未来的关键

智能产业发展离不开科技创新。

据2021年第四季度的数据，重庆在智能产业领域已建成国家级企业技术创新中心21家，布局研发机构的企业约1500家。中国信息通信研究院西部分院等一批高端研发检测载体先后落户。联合微电子中心、英特尔FPGA中国创新中心等一批高端创新平台也已建成投用。

同时，重庆陆续引进了北京大学、清华大学、中国科学院大学、上海交通大学、哈尔滨工业大学、华中科技大学、湖南大学、电子科技大学、新加坡国立大学等众多中外知名大学。这些知名大学以研究院、研究中心、科创中心等不同形式落户重庆，增强了重庆科技创新力量，弥补了重庆人才资源短板。

据 2022 年 1 月的消息，科技部支持重庆建设国家科技成果转移转化示范区，这是"十四五"以来科技部批复的第 3 家国家科技成果转移转化示范区。基于示范区框架，重庆将通过加速打造科技成果转化体制机制、科技成果转化服务体系，发展环大学创新生态圈，打造高水平创业孵化平台，打造科技成果区域协同转化等手段，优化科技成果的源头供给，提升科技成果中试熟化水平，加速产业迭代升级。

这是重庆智能产业未来的创新驱动力。

此外，重庆角逐智能未来的关键还在于数字化基建情况及基于优势产业的顶层设计。

据《科技日报》报道："重庆以 5G 为发力点，积极布局物联网创新发展，夯实工业网络基础。重庆成为全国首批 5G 规模组网试点城

● 工业互联网标识解析国家顶级节点（重庆）
Top nodes of the National Industrial Internet (Chongqing)

市，累计建成 5.3 万个 5G 基站，并构建 5G 产业体系；逐步完善国家工业互联网标识解析顶级节点（重庆）基础设施，标识注册量和累计解析量同比增长位列全国第一；承接国家'火星·链网'区块链超级节点；建立了 10 个核心大数据运营商在内的服务体系及 20 个大型云平台，算力水平西部领先。"[1]

5G 基站的建设、工业互联网标识解析顶级节点的上线、云平台的运算能力，直接决定一个地区智能工厂的数据共享能力、传输能力与应用能力。

而基于优势产业和新兴产业的顶层设计，为未来的智能产业发展指出了着力点。这些顶层设计主要体现在 2021 年 8 月发布的《重庆市制造业高质量发展"十四五"规划（2021—2025 年）》（以下简称《规划》）中。

《规划》提出，重庆将不断增强电子、汽摩、装备制造、消费品、原材料等支柱产业的国际竞争力，并将重点建设一批具有全国影响力的战略性新兴产业集群，涉及新一代信息技术、新能源及智能网联汽车、高端装备、新材料、生物技术和绿色环保 6 类产业。

其中，在新一代信息技术方面，面向"智造重镇""智慧名城"建设需求，建设重要的功率半导体器件、柔性超高清显示、新型智能终端、先进传感器及智能仪器仪表、网络安全产业基地和中国软件名城。

新能源及智能网联汽车方面，发挥在燃油汽车生产能力、零部件配套体系和集成电路生态等方面综合优势，促进新能源汽车与信息通信、能源、交通深度融合，建设国内领先的动力电池产业基地、氢燃料电池应用示范基地和国内先进的汽车电子产业基地。

高端装备方面，推动传感器、通信模组等组件在整机中植入有色

1　雍黎，科技日报，《点面结合提升制造业 重庆向"智造重镇"演进》，2021 年 8 月 30 日。

● 2021 智博会上，氢燃料电池汽车产业集中签约仪式
At the 2021 Smart China Expo, a centralized signing ceremony for the hydrogen fuel cell vehicle industry

合金、合成材料等新材料应用，建设国家重要的中高端数控机床、城市轨道交通车辆、新能源装备产业基地和西部领先的工业机器人、增材制造装备产业基地。

新材料方面，面向产业发展和重大工程建设迫切需求，建设具有国际竞争力的聚酰胺材料、聚氨酯材料产业基地和国家重要的先进有色合金材料、玻璃纤维及复合材料产业基地。

生物技术方面，面向重大疾病发现和居民健康管理领域需求，加快生物药上市步伐，推动医疗器械、化学药原料药及制剂、现代中药升级发展，建设国内一流的生物医药产业集聚区。

可以说，战略性新兴产业集群的发展规划是建立在原有支柱产业上的一次更新、更智能的进化，而原有的支柱产业也离不开智能化赋能，两者相辅相成，从而形成更有集群效应、更富产业链价值的优势产业。

2021 智博会主论坛上，全国政协副主席、中国科学技术协会主席万钢在演讲中透露，2020年中国人工智能产业规模已达到3030亿元，同比增长15%，高于全球的增速，最后更是对重庆寄予厚望，希望重庆成为内陆的智能产业和数字经济发展高地，高水平国际开放合作的聚集地。

过去5年，通过智博会的引"智"引"实"，通过智能时代的基础设施建设、创新的激发与成果的转化、原有产业的"智变"与新兴产业的打造以及清晰的路径规划和强有力的执行，作为智能产业和数字经济发展高地的重庆，作为高水平国际开放合作的聚集地的重庆，正在到来。

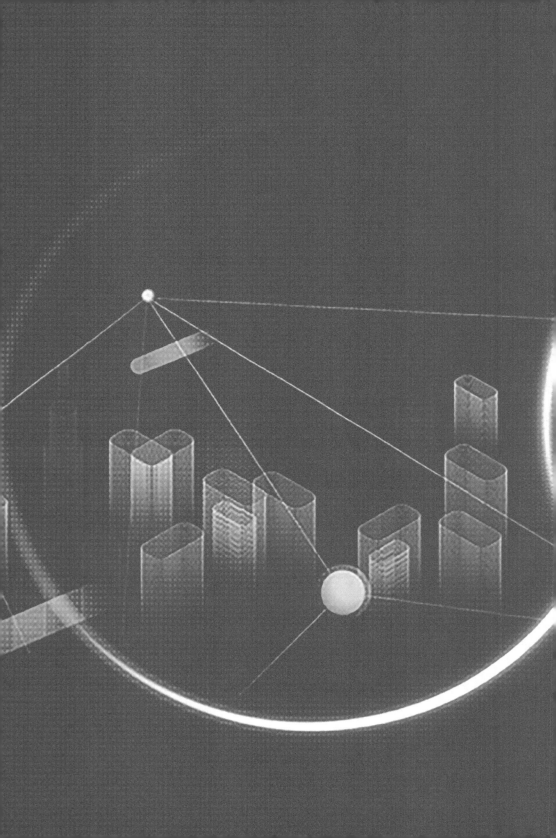

第三章

智慧生活：为生活添彩的智能化内核

　　数字的逻辑融入社会，智慧的算法铺满城市，智能化生活的内核一旦启动，数字文明也就融入了人们对生活的美好想象，于是，"智能办公""智能购物""智能家居""智能医疗"……整个城市的生活变得更加多姿多彩。

　　无论城市怎样进化，无论技术如何变革，智慧城市的主角，永远是生活其中的市民。

第一节
智能化与数字化，
已经渗透人类生活每一个角落

网络上偶尔会有人提出一种假设：如果没有人工智能，人们的生活将会怎样？

失去人工智能，城市交通将会瘫痪，无法网络叫车，公共交通运行受阻；失去人工智能，商业物资运转不畅，市场供需急剧失衡，人们购买生活物资不再便捷；失去人工智能，无法移动支付，人们重回现金时代，很多依托于移动支付的全新生活方式将消失殆尽……

人工智能已经与人类生活的方方面面紧密地联系在一起，无论是经济发展、城市治理，还是民生生活，都离不开人工智能。

我们不难感受科技发展和数字化的到来，不断更新换代的智能产品涌入我们的生活之中，而这些智能技术给我们生活带来的便捷体验，正在以肉眼可见的速度逐年提升。"智能办公""智能购物""智能居家""智能医疗"，大众生活的方方面面，因为人工智能的底层支撑而变得更好。

智能化、数字化的不断演变，已经成为影响人类社会发展的重要力量，并渗透到人类生活的每一个角落。

"智慧出行"，缔造智能时代新格局

从两腿跋涉到借助驴马，古代人类借外力避免了出行劳苦；从蒸汽动力到燃油动力，近代工业以能源提升了出行速度；及至科技飞速发展的当代，飞机、高铁更是可以让人千里之外，朝发夕至。科技的进步，大大降低了人类的体力损耗，并大幅提升了出行速度。

然而在动辄人口上千万的现代都市，超高密度的出行需求、超高速度的生活节奏，正在对人类出行提出更高的要求：降低时间损耗、减少彼此干扰。

在这种全新的需求下，人工智能提供了最佳的出行解决方案。

清晨上班，不用急匆匆地赶公交，可以一边吃早餐一边通过 app 关注公交车到站信息，从容出门准时踏上所需车辆；开车时，不用为道路拥堵而担心，智能导航实时更新路况信息，帮助规划出行线路……这些出行的智能体验，已经成为人们的生活日常。

● 2021 智博会上展出的智慧交通场景
The scenario of smart transportation exhibited at the 2021 Smart Chian Expo

作为城市发展的主要动力，交通对生产要素的流动、城镇体系的发展有着重要的影响，是城市兴衰的决定性因素之一。在人工智能飞速发展的时代，人们已在各类出行生活场景中享受到智慧科技所带来的便利。

在 2021 智博会的华为展台，主题是"把数字世界带入每个人、每个家庭、每个组织，构建万物互联的智能世界"。在智慧新出行方面，华为智慧机场、智慧城轨等大交通场景化解决方案依次亮相智博会展台，以"运控一张图"、"出行一张脸"、"机场智慧大脑"、城轨云平台等数字化、智能化解决方案，为参观者勾勒出未来智慧出行的一个个新场景。

而在现实应用中，重庆机场着力打造的"智慧机场"生态圈被誉

● 华夏云融航空科技 5G 智慧机场数字平台
China Express Yunrong Aviation Technology 5G Smart Airport Digital Platform

● 2021 智博会长安汽车展位
Chang'an Automobile booth at the 2021 Smart China Expo

为重庆智慧城市建设的"样板支撑"。在这里，旅客从值机、安检到登机，实现了全流程"无纸化"乘机，9 秒国际自助通关[1]，"一张脸走遍机场"。此外，作为全国首批试点的 6 大机场之一，重庆机场还上线了国航执飞的"首都 T3 －重庆"航线行李全流程跟踪查询，开启了行李数据共享服务。

在智博会秀出智慧出行"绝活"的还有长安汽车。长安汽车董事长朱华荣在智博会高峰论坛上，向各界描绘了企业关于"新汽车新生态"的战略布局。而在其展位，长安汽车发布了多项智慧出行"黑科技"，比如：APA7.0 远程无人代客泊车系统，可以为车主打造"呼之即来，挥之即去"的专属停车场"私人代驾"；全球首发的电驱高

---

1 黎静，上游新闻，《打造"智慧机场"，重庆机场要这样干》，2021 年 4 月 15 日。

频脉冲加热技术，即便在 –30℃极寒环境下，也可以实现 5 分钟加热20℃，提升电车的低温行驶性能，配合余热回收技术，将续航里程提升 40 ～ 70 公里。

"新居住"时代为生活添彩

智能家居在中国的发展历程，经过了"萌芽－开创－徘徊－演变－爆发"几个阶段。

90 年代的萌芽期，智能家居还处在一个概念熟悉、产品认知的阶段，这时还没有出现专业的智能家居生产厂商；从 2000 年开始的开创期和徘徊期，深圳上海等一线城市开始出现完善的智能家居市场营销、技术培训体系等；从 2011 年开始的演变期，智能家居产业的放量增长进入拐点；直至 2020 年之后，智能家居产业开始进入爆发期。

目前，随着智能单品逐步落地，中国的智能家居市场容量和渗透率正处于提升期。可以说，智能家居的研发、设计、生产、销售等上下游行业已经吸引了众多玩家入场，支撑起一条完整的产业链，让智能家居产业得以大规模快速发展。人工智能、物联网、云平台等技术的发展开始为智能家居产业铺平道路。在技术研发和消费需求的双重驱动下，智能家居已经演变为全屋智能的"新居住"要求。

不同于传统居住，"新居住"重在全场景体验，即全屋智能——根据所有居住场景，为用户提供覆盖全部需求的解决方案，以及厨房、卧室、浴室等各空间的局部焕新及整装方案。

中国家庭对智能家居的兴趣已经支撑起了一个庞大的市场。中研普华研究院发布的《2022—2027 年智能家居产业深度调研及未来发展现状趋势预测报告》认为，中国将占据全球 50% ～ 60% 的智能家居市场消费份额，成为全球最大的智能家居市场消费国。中商产业研

解码智能时代
从中国国际智能产业博览会瞭望全球智能产业（2018－2022）

图片来源：张锦辉/视觉重庆
Photo by: Zhang Jinhui/Visual Chongqing

● 2021 智博会期间，在礼嘉智慧公园智慧生活馆，智慧厨房岛台能指导参观者烹饪重庆小面

At the 2021 Smart China Expo, in the Smart Life Pavilion of Lijia Smart Park, the smart kitchen island guided visitors to cook Chongqing noodles.

究院预测，2022 年我国智能家居市场规模可达 6515.6 亿元。[1]

据天眼查专业版数据显示，2021 年国内存在近 16 万家智能家居相关企业。除了批发、零售型企业，从事智能家居方面的信息传输、软件和信息技术服务的企业超过 2.5 万家，为智能家居提供科学研究、技术服务的企业也有近 2 万家。[2]

事实上，人们对"家"已经有了更细致的审视——对住宅的健康属性愈加重视，对社区的运营管理更加关注，对空间的功能需求不断提高。而细化到家居产品，功能上不仅要更舒适，还要更健康、更"聪明"。

在 2021 智博会现场，各厂商都拿出了智能家居的绝活。在战略、技术、方案、应用落地等方面"秀肌肉"。力求基于行业最大的智慧家庭场景生态，把消费者带入"新居住"时代。

1 中研普华研究院，《2022—2027 年智能家居产业深度调研及未来发展现状趋势预测报告》。
2 界面新闻，《天眼查：我国现有近 16 万家智能家居相关企业，2020 年新增超 3 万家》。

● 2021 重庆市大数据智能化应用十大 "智慧社区" 精选案例
2021 Top Ten "Smart Community Service" Selected Cases of Chongqing Big Data Intelligent Application

过去的智慧家居被广泛地聚焦为室内场景的物联网应用，而在2021年智博会上，智能家居呈现出更为体系化的智能化场景应用解决方案，智慧场景也从家庭应用扩大到社区应用，从智慧家居延伸到智慧社区。

在智慧家居方面，OPPO 也基于 UWB 空间感知技术，带来了精准指向控制智能家居的全新体验，让智能设备之间的控制更为精准；中国移动的 "智慧家庭" 展区，通过打造 "一张全光网络 N 种标准产品 + 多应用场景" 创新的产品体系，可一键定制全屋场景化模式，让观众感受家居生活 "更智能、更安全、更环保、更有趣" 的体验。

而在智慧社区方面，金科·博翠未来、美的万麓府、龙湖智慧服务、重庆万科物业等 5 个智慧小区和 5 家物业企业，以各具特色的智能化场景应用，从日常生活的方方面面，为居民生活带来了便捷高效的居住体验，有效提升了物业管理服务的效能，居民的获得感、幸福感、安全感不断跃升，因此被评为"2021重庆市大数据智能化应用十大'智慧社区'精选案例"。

"数字化"构筑美好生活新图景

人类的科技和文明是建立在美好生活之上的，而美好生活所延伸出来的美好智慧生活，正在由万物互联、万物可感知的世界书写。我们所处的世界正在迅速、深度地步入数字化时代。

重庆担负着国家数字经济创新发展试验区和国家新一代人工智能创新发展试验区建设的双重使命。近年来，重庆围绕制约数字经济和新一代人工智能创新发展的关键问题，努力构建新型的基础设施和保障体系、政策支撑体系，大力开展科技创新、改革试点、应用示范，积极推动产业培育和经济转型，正在成长为内陆智能产业和数字经济发展的高地、高水平国际开放合作的聚集地。

连续举办4届的智博会，也已经成为重庆展示数字经济发展的常态化名片。在具备当代都市的典型特征之外，重庆也是了解中国现代数字化城市发展的重要窗口。

在2021智博会上，全国政协副主席、中国科学技术协会主席万钢表示，人工智能的行业应用广度在日益扩展，其应用正在走出实验室，进入制造、交通、商贸、法律、销售、服务以及城市管理等生产、生活的场景，进一步发挥赋能作用，有力推动了各个领域的智能化、数字化、网络化发展态势。

在过去的两年多时间里，新冠疫情迫使很多事不得不按下"暂停键"，但数字信息技术却插上了"腾飞之翅"，线上授课、线上办公、线上购物等将我们带进了更智能的时代。

智能服务是数字经济为生活带来的直接改变，数字化生活正在颠覆着传统生活方式。云拜年、云学习、云办公、云旅游，万物皆可"云"成了数字生活的标配。

一时之间，数字化转型成为千行百业的刚需，不仅被认为将带来生态价值、经济价值的提升，更会成为推动整个人类社会更智慧、更高效、更高质量发展的关键布局。

重庆新型智慧城市运行管理中心已接入 200 多个系统，重点打造近 30 个综合应用场景。"渝快办""渝快政""渝快融""渝康码"等智能化创新应用，正在为重庆市民的城市生活、生产经营带来巨大便利，让城市管理更加智慧。

智能门禁、智慧厨房、智慧节水、智能温控、光线调节等一体化智能家居，正为居家生活添彩。

刷脸 / 二维码入店、可视化搜索商品信息、感应区结账等功能应用在无人售卖商街，让外出购物更加便捷。

网上支付、人脸识别、景区智能导服等技术的应用，为出门旅游提供更便捷、更强大、更人性化的智能服务。

"衣食住行游购娱"中的智慧化应用，不断满足着人民群众高品质生活需要。[1]

让数字化、智能化技术走入人类生活、让人类生活得更加美好，是科技与人类共处的理想图景，也是重庆的愿景。

1　重庆日报，《构建智慧名城"四梁八柱"重庆智慧城市建设水平走在全国前列》，2021 年 8 月 18 日。

第二节
数字化新生活，
中国生活方式正在引领全球潮流

城市的数字化、智能化、可持续发展，已经成为构建人类命运共同体中极其重要的一个议题。

近年来，中国围绕数字科技、产业数字化、乡村振兴、碳达峰与碳中和、医疗与健康、数字生活与服务等大领域，努力建设一个具有数字经济新优势的中国。

中国的数字化生活正在引领全球潮流，进一步向更广阔的天地行进。

重庆：打造"数字生活"新方式

2018 年至今，每年一度的智博会，已经成为外界关注智慧城市最新进展与发展趋势所不能错过的风向标。

通过智博会，我们可以思考：经历十余年发展的智慧城市产业是否已经来到了全新阶段？如何让智能化数字化技术如江水一样滋养城市发展？

事实上，经过十余年的努力，重庆不断加快传统产业数字化、智能化与实体经济融合发展，已经建设成为中国的数字经济示范区。

2021 年 12 月 8 日，重庆市政府印发《重庆市数字经济"十四五"

发展规划（2021—2025 年）》提出，到 2022 年，全市将集聚 100 家数字经济龙头企业、500 家前沿领域高成长创新企业、5000 家"专特精尖"中小微企业和创新团队，创建 10 个国家级数字经济应用示范高地，到 2025 年全市数字经济总量超过 1 万亿元。

具体实施上，重庆瞄准大数据、人工智能、集成电路、智能网联汽车等智能产业重点发展领域，倾力打造"智造重镇"，有效推动产业层级跃升和经济高质量发展；以大数据智能化推进城市治理提升，着力建设"智慧名城"，政务服务、社会治理、百姓生活变得更加智能、绿色和高效。

在重庆，基于互联网、大数据的"智慧交通"使城区拥堵情况大为缓解；"智慧防涝"大幅提高了城区内河调蓄效益和排水防涝应急处置效率……甚至，数字化生活方式已经进入普通百姓家庭，解决了许多民生"痛点""堵点""难点"问题。

重庆便民服务平台"渝快办"，在 2021 年迎来了 3.0 版本。在便民服务贴心化方面，实现全市 42 个政务服务大厅在 app 上的"一键查找""一键导航"；同时，在建设银行重庆市分行 277 个营业网点设立劳动者港湾，为户外劳动者歇脚休憩提供饮水、充电、无线上网等便利服务。

此外，"渝快办"3.0 有针对性地面向不同群体推出了不同的服务应用。如针对老年人群体开设服务专区，上线社保、医疗、公积金、户政、证明服务 5 大类 20 余个事项，同时新增操作提示、语音辅助等功能，提供大字号、大图标、高对比度以及更为简洁的界面风格，带来多样化的暖心服务。[1]

在新冠疫情仍然没有完全消散的当下，渝康码一直在为 3000 多

---

1 刘巧、陈诗宜，新华网重庆，《重庆网上政务服务平台"渝快办"3.0 正式上线》，2021 年 7 月 28 日。

万重庆人民保驾护航。其背后的数字力量是浪潮云安全可靠的技术支撑。小小的二维码背后有着体量极为庞大的数据支撑，包括每个人的健康数据、卫健疾控部门提供的就诊信息等。

重庆深知，决定城市发展的将不再是土地和人口红利，而是基于大数据运营和服务所产生的数据红利，这将基于数据资源形成各种智慧应用，推动智慧城市建设。而重庆的数字化潜能远比你我想象的更加强大。

中国数字化生活引领全球

"数字重庆"建设的显著成就正是"数字中国"建设快速推进的缩影。伴随着数字产业化和产业数字化"双轮驱动"、实体经济和数字经济相互融合，中国数字经济气象万千。

据国家互联网信息办公室发布的《数字中国发展报告（2020年）》显示，"十三五"时期，我国数字经济发展活力不断增强，我国数字经济总量跃居世界第二，数字经济核心产业增加值占 GDP 的比重达到 7.8%。[1]

"数字化"在中国的迅速普及与国人对数字化生活的开放态度密切相关。以网络支付为例，早在 2017 年中国就已经成为全球最大的数字支付市场，二维码已成为广泛使用的支付手段。

2018 年 3 月 5 日，中国人民银行公布了《2017 年支付体系运行总体情况》，数据显示，2017 年移动支付业务量保持较快增长，网上支付业务 485.78 亿笔，金额达 2075.09 万亿元，移动支付业务 375.52 亿笔，金额达 202.93 万亿元。

---

1 张阳，环球网，《〈数字中国发展报告（2020 年）〉发布 我国数字经济总量跃居世界第二》，2021 年 4 月 26 日。

短短四年之后，2022 年 4 月 2 日，在中国人民银行发布的《2021 年支付体系运行总体情况》中，2021 年网上支付业务 1022.78 亿笔，金额 2353.96 万亿元，移动支付业务 1512.28 亿笔，金额 526.98 万亿元。

日常生活的深度数字化，使中国消费者网上支付业务笔数比起四年前翻了一番，金额稳步增长，而移动支付笔数增长到 2017 年的 4 倍，移动支付金额也达到了四年前的 2.6 倍。

值得注意的是，即使是全球消费受到疫情的影响，也难撼动中国在世界数字经济发展中的地位。2021 年 8 月 2 日，中国信息通信研究院发布的《全球数字经济白皮书——疫情冲击下的复苏新曙光》显示，2020 年中国数字经济规模为 5.4 万亿美元，位居世界第二。[1]

疫情之下，"宅"成为多数人的日常。需求在哪里，供给就朝向哪里。海量的人群"足不出门户、行不逾小区"，庞大的生活消费需求叠加疫情防控的严峻形势，既要减少交叉感染的风险，又要兼顾复工复产的现实需要，于是出现了"无接触商业"。

以餐饮业为例，中国饭店协会曾与多家餐饮品牌联合落地首批"无接触餐厅"，一些大型商场和超市着力发展线上销售，各大快递公司出台各自的无接触配送标准……通过线上交易、线下定点配送、用户自提等方式，避免买卖双方直接接触，这种"无接触商业"模式受到消费者青睐。[2]

"无接触商业"除了解决疫情之中大众的生活所需，也倒逼数字经济本身萌发出各种全新的商业模式，各种新商业、新服务层出不穷，比如无人货架、无人超市、智能快递柜、无人配送车等，很大程度上改变了零售的业态，也创造出了新的就业机会。

1　中国信息通信研究院，《全球数字经济白皮书——疫情冲击下的复苏新曙光》。

2　郑志辉，新快报，《"无接触餐厅"来了！中饭协、美团联合多家餐饮企业力推全链条"无接触"》2020 年 2 月 21 日。

图片来源：张锦辉／视觉重庆
Photo by: Zhang Jinhui／Visual Chongqing

● 在礼嘉智慧公园，人们利用路过的无人售货机选购饮品
People buy drinks from passing vending machines in Lijia Wisdom Park.

此外，"云旅游""线上读书会""影院卖品外卖"等文旅新品应运而生。事实上，我们的生活早已离不开数字技术。我们不仅早已"在场"，而且时时"在线"。"隔离疫情不隔离经济"，不得不说，中国的数字化生活创新，是引领经济社会发展和影响国际竞争格局的重要力量。

2022 年 2 月 5 日，时任巴基斯坦总理伊姆兰·汗（Imran Khan）正式访问中国并出席北京冬奥会开幕式，他表示："以抗击新冠肺炎疫情为例，环顾整个世界，没有任何一个国家可以像中国这般采取有力的应对措施。"[1]

1　邢晓婧、刘彩玉，环球时报，《巴基斯坦总理伊姆兰·汗接受〈环球时报〉专访："自信的国家，才能举办这场盛会"》，2022 年 2 月 7 日。

外界眼中的中国数字化

除了带动中国经济转型发展，"数字化"还在提升中国社会管理效率、推进社会发展进程方面扮演着重要角色。经过这几年的跨越式发展，"数字中国"已经不仅仅是一个经济概念，而是渗透在社会发展的每一个环节中。

数字化浪潮下，各种新业态、新服务、新模式不断涌现，自主创业、副业创新、灵活用工等生机勃发。在农村，一部手机、一根自拍杆，正成为越来越多农民的"新农具"，把农产品和消费市场直接链接在一起；在城市，网约车、外送服务奔波在大街小巷……数字中国的探索与创新，正深入国计民生的每一个角落，让创新创业的活力迸发，让便民利民的红利涌流。

"数字化"已成为提升国人生活水平、优化经济结构、推动社会发展的重要途径，中国也成为全球数字化发展进程最快的国家之一。

而众多海外媒体对中国在数字科技发展和数字经济领域取得的成就都给予了充分肯定。论及中国的数字化，美国《财富》杂志曾经这样写道："中国已经进入一个数字化产业的新时代，未来数十年，中国将成为全球数字化发展的引领者。"

2022 年 4 月 13 日，在北京大学与芝加哥大学联合系列讲座的高端对话中，麦肯锡公司全球资深董事合伙人、麦肯锡全球研究院院长华强森（Jonathan Woetzel）指出，中国拥有独特而成熟的数字生态系统、深厚的数字消费者基础和飞快的数字变革发展速度。通过一系列数据分析和图表演示，华强森博士揭示了中国在全球数字变革中占据的领先位置，中国数字变革的开拓者们从高频服务中脱颖而出并扩展到全方位收购服务。[1]

1　丁舒疃、谢智愚，北京大学官方网站，《麦肯锡高管华强森、成政珉谈"中国数字变革及亚洲影响"》，2022 年 4 月 13 日。

可以看到，数字经济大潮正在中国如火如荼地展开，中国"数字化"作为一种新兴经济发展模式，已经获得国际社会的广泛认同，甚至成为了耳熟能详的"中国名片"。

第三节
建设"智慧名城"，
重庆成为智慧、乐业、宜居城市

城市，承载着人类对美好生活的向往。它的每一次进化升级，都是人类社会进步的表现。

一座城市的成长和科技的进步紧密相关。随着新基建的加速推进，围绕技术、政策、生态加速构建"智慧城市"，正在成为每一个经历科技革命洗礼的城市的共同命题。

新型智慧城市建设如火如荼。这些年，重庆矢志不移地推进智能产业的规划与发展。这座山水之城不仅将互联网、物联网、云计算等现代科技融进了"智慧城市"的概念中，更将能感知、会思考、会进化、有温度的"梦中之城"带进了我们的现实生活。

"智慧名城"的进阶之路

2021 年，被称为重庆加速发展数字经济、提速"智慧名城"建设的关键年。

2021 年 6 月 15 日，《重庆市城市基础设施"十四五"发展规划（2021—2025 年）》提出，到 2025 年，将构建起高效实用、智能绿色、安全可靠的现代化信息基础设施体系，为实现全市经济社会向更

解码智能时代 从中国国际智能产业博览会瞭望全球智能产业（2018—2022）

● 2021 智博会忽米网展台
Humi Network booth at the 2021 Smart China Expo

高质量发展、加快建设现代经济体系提供强有力支撑，助力"智造重镇""智慧名城"建设。

早在 2019 智博会上，重庆就响亮地提出，集中力量建设"智造重镇"和"智慧名城"，在数字产业变革中驭势而为，掌握新一轮科技革命主动权。[1]

这种旗帜鲜明的集中力量，既兼顾了重庆这座老牌工业基地的深厚积淀，也锁定了智能时代数字经济发展的未来方向，为重庆市的长远发展规划出了清晰的路径。

只要你在智博会展馆里游览一圈，就不难发现，重庆的传统制造业正在焕发新生，与工业互联网平台深度联动的智能制造，已经无处不在。

截至 2021 年 8 月，重庆建成 200 个智慧农业示范基地，累计推

---

1 韩政、陈翔，重庆晨报，《重庆智能产业发展已处"第一方阵"下一步目标建智造重镇、智慧名城》，2019 年 8 月 30 日。

动 350 所智慧校园、99 个示范智慧旅游景区示范建设，建成智慧小区 191 个、智能物业小区 545 个，已建成 44 家智慧医院、初步建成全市健康医疗大数据资源湖和平台基础，还重点打造应急管理、智慧消防、村社法律服务等基层治理应用场景，以及智慧停车、明厨亮灶、智慧交通等民生领域应用场景，智慧生活正在进入市民生活的方方面面，带来翻天覆地的美好变化。[1]

北碚区建成投用重庆市首个智慧未成年人保护系统，将困境儿童分为 3 类，为部分存在监护缺失等较大安全风险的儿童配备了智能腕表，监护人、社工、社区可以通过语音通话、健康数据监测、跟踪定位等功能，及时掌握儿童状况，并为紧急救助提供精确定位，降低未成年人可能面临的危险。

重庆首个智慧监管农贸市场——半岛逸景农贸市场，近百个摊位按鲜、活、生、熟、干、湿等区域整齐划分，各区域大屏幕滚动公示当日蔬果抽检结果，每个摊贩面前立着一块电子屏，经营者个人信息、星级评分等级、当日菜价等信息一清二楚。[2]

在重庆高新区一条里程 5.4 公里的环形路线上，自动驾驶巴士接驳服务已于 2021 年 8 月底正式运营，该项目通过 C-V2X、5G、边缘计算等技术，以车路协同为核心理念，建设了智慧公交站、智慧斑马线、智慧匝道、智慧十字路口等智能化设施设备。无人驾驶车辆在行驶中，不仅能够平稳地进行加减速、变向等操作，遇到障碍及车辆能够进行避让，还能与其他社会车辆"融洽相处"，此外，车辆内屏可显示车身周围物体信息和车辆预计行驶的路线，科技感十足。据了解，该项目运营模式还被应用到高新区西永、大学城等区域，有效解决民众出行问题，缓解交通拥堵压力。

重庆南岸区长嘉汇购物公园前，由"物联网＋智能传感"技术

---

1 重庆日报，《让生活更精彩！"智慧名城"建设渐入佳境》，2021 年 8 月 20 日。
2 罗斌，《重庆日报》，《城市更聪明、市民更舒心 智慧赋能重庆市民幸福生活》，2022 年 7 月 2 日。

升级的"智能斑马线"项目，已于 2021 年 10 月正式"上岗"，项目集人流监测、闪烁道灯于一体，通过安全岛改造、增设地面发光道钉、发光警示带以及人流监测仪，双向提醒来往车辆、行人注意交通安全。

在重庆市两路寸滩保税港区，重庆智慧物流应用的一个新场景已经落地。重庆飞力达立体仓库，实现了影像识别－入库－出库－AGV 搬运车全系统的智能化和无人化，自动化仓库不仅存放货物容量和入库速度均是传统仓库的 3 倍，出库速度也是传统仓库的 2.5 倍，出入库速度可达每小时 960 箱，准确率达 100%。[1]

用智慧之光照亮城市发展的未来，除了遍布重庆市的主城区，辖区内各区县也在加快打造智慧城市的步伐。重庆市永川区政府在智博会上介绍，永川为发展智慧交通，实现了当地金龙客车与百度联合研发的 L4G 自动驾驶中巴车首发；为打造智慧城管，永川接入了视频监控系统，建立感知、分析、服务、指挥、监察五位一体的智慧管理体系；此外，永川区还与航天科工达瓦科技联合开发数字孪生城市决策平台，为城市建管、空间规划、消防救援等提供决策辅助。

让"近者悦、远者来"，重庆的"乐业基因"

2000 多年前，亚里士多德曾说："人们为了生活而来到城市，为了生活得更好而留在城市。"今天，大数据智能化为城市品质的提升注入了更多"智慧因子"，让智能化精准对接民众的美好生活需求。

宜业，是一座城市孜孜以求的美好愿景，更是一座城市最长久最旺盛的生命力。一个能留住人才，并让其心甘情愿扎根的城市，一定

---

1 顾立、张韵秋、冉黎黎、谭力，上游新闻，《智汇八方丨重庆飞力达，为智慧物流提速》，2020 年 11 月 18 日。

是有着"乐业"基因的，比如悠久的历史、良好的经济环境、便利的交通、醇厚的文化底蕴、利好的人才政策等。

最年轻的直辖市、新一线网红城市、8D 魔幻之城……重庆，这个带着多重标签的山水之城，这个有着超过两百万外来人口的城市，如何让外地人安心进城、稳定就业？如何更好地服务经济社会发展？如何改善民生让"近者悦、远者来"？

一座重庆城，傍水而成、依山而建，原本既不利于现代交通的发展，也有悖于现代建筑的规划，然而在直辖之后，经济发展强劲，人口不断聚集，以极快的速度成长为一座国际大都市，重庆这座城市已经创造了太多的奇迹。

● 成渝地区双城经济圈工业互联网产业投资基金启动战略合作
Chengdu-Chongqing Region Twin Cities Economic Circle Industrial Internet Industry Investment Fund launched strategic cooperation.

交通方面，重庆 2021 年按下"快进键"，用"加速度"跑出了"高效率"：长江上游首个万吨级码头重庆新生港开港营运；成渝地区双城经济圈战略性、标志性、示范性重大项目——设计时速 350 公里的成渝中线高铁启动建设；合川至长寿高速即将建成通车，重庆主城都市区将进入"三环时代"；城口到开州高速开州谭家至赵家段正式通车，重庆"县县通高速"指日可待……

经济方面，根据 2022 年 3 月 18 日，由重庆市统计局、国家统计局重庆调查总队发布的《2021 年重庆市国民经济和社会发展统计公报》显示，2021 年，重庆市地区生产总值两年平均增速 6.1%，在长江经济带 11 省市中居第 3 位，社会消费品零售总额两年平均增速居第 2 位，城镇和农村常住居民人均可支配收入两年平均增速均居第 1 位，重庆在长江经济带的支撑作用不断提升。2021 年，重庆全年主城都市区、渝东北三峡库区城镇群、渝东南武陵山区城镇群分别实现地区生产总值 21455.64 亿元、4895.15 亿元和 1543.19 亿元，分别比上年增长 8.0%、9.1% 和 7.6%。[1]

人才政策方面，2021 年重庆市人才引进工作采取"走出去、请进来"的举措。其中，2021 年"百万英才兴重庆"北京行、四川行等活动，就是重庆人才引进工作"走出去"的最新要点之一。而针对人才"请进来"的管理等工作，重庆人力社保部门坚持分类推进人才评价，建立人才分类评价职称评审"绿色通道"、科研项目经费"包干制"、专家"揭榜挂帅"等，有力激励人才脱颖而出。据了解，2021 年重庆人才服务已升级至 3.0 版本，可通过电话、网站、微信、平台四种方式为重庆英才卡持卡人才提供 25 项服务。

……

2022 年 5 月 11 日，在更为清晰的城市战略指引下，重庆市聚焦

---

1 赵颖竹，华龙网，《经济总量迈上 2.7 万亿元台阶！2021 年重庆经济成绩单藏着这些亮点》，2022 年 3 月 18 日。

重点产业和重点领域，面向全球启动了精准引进急缺工程类人才的行动。具体而言，就是围绕重庆市 7 大支柱产业和 33 条产业链、国家级专精特新"小巨人"企业等，通过"线上 + 线下"相结合的方式，组织针对性的引才活动，引进海外工程师和卓越工程师。

在世界热切的目光中，承载着领跑长江上游经济的重庆已经站在了新的起点，开始了改写历史的超超征程：一幕幕现代大都市崛起的大戏在这里开启，一条条通往智能产业时代的大道在这里建成。

打造"重庆式幸福生活"

无论是本地人还是外地人，爱一个城市自有其充足的理由。对于普通百姓来说，愿意留在一座城市的理由，除了经济是否有活力、能否提供充足的就业岗位等物质保障外，美好的精神生活追求也成为重要的考虑因素。

如何判定一座城市是否宜居？环境优美、社会安全、文明进步、生活舒适、经济和谐、美誉度高，构成了宜居城市的基本特征。

海纳百川、开放包容；城在山上建、江从城中流，水在城中、人在景中；峡谷魂、巴山夜雨、云海仙境；洪崖洞灯火璀璨、山城巷人来人往；清晨的小面、夜晚的火锅……

在重庆，你能感受到历史与现代、传统与时尚的碰撞，也能感受到重庆人刻在骨子里的坚毅和豪爽。不论从气候、城市建设、美食、风景、环境来说，重庆都不失为一座宜居的城市。这片神奇的土地，也被外地游客誉为历史之城、文化之城、旅游之城和文明之城。

开放的政策、包容的城市气质，诗与远方都在眼前。这让越来越多的年轻人选择在这里工作、生活，愿意在这里追梦、造梦。

有梦追就很幸福，幸福的人各有各的体验，幸福感弥漫的生活始

解码智能时代 从中国国际智能产业博览会瞭望全球智能产业（2018—2022）

● 以拥抱智能时代的方式，打造重庆式幸福生活
Build a Chongqing-style happy life by embracing the intelligent era.

终令人向往。而当一座城市在面向未来的时候，拥有一个共同的、清晰的、宏大的梦想，人们置身其中，幸福感就拥有了更加广阔的承载空间。

城市建设归根到底是为了生活在其中的人，只有让市民享受数字化生活的便利、分享智能化时代的红利，这座城市才拥有更美好的未来。

五年智博会的持续主办，不但为重庆的智能产业注入了极强的创新力与竞争力，也为重庆的广大居民带来了满满的获得感与幸福感。

近年来，重庆一方面布局产业数字化、数字产业化，推动传统产业转型升级，培育战略性新兴产业，实现经济高质量发展；一方面抓人居环境、教育、医疗等民生事业，利用智能化发展成果，反哺城市建设，提高民生福祉。高质量发展、高品质生活，"两高"并进，为重庆人建设了一条通往幸福生活的"高速路"。智博会便是用这样的方式宣告，重庆正以打造宜居、韧性、智能的现代化城市为目标，努

● 智博会的永久落户，带给重庆居民时代红利

The continuous hosting of the Smart China Expo will enable Chongqing residents to share the dividends of the times.

力让人民群众在城市生活得更方便、更舒适、更美好。

事实上，重庆快速的成长不断聚合着西南地区各种要素和资源，使这个年轻的直辖市一步步变成一座现代化的大都市；永久落户重庆的智博会，更让世界看见了重庆。

而"城市的宏大"与"城市的温度"并不对立，站在"十四五"的开端，重庆正以独有的城市特质、人文精神，向全世界展示"重庆式幸福生活"。

第四节
五年日积月累，重庆以智能化为生活添彩

如同人一样，城市也需要在各方面不断成长和成熟。

评判一座城市的发展水平，除了经济增长、基础设施、城市治理等常规指标，在一个全新到来的智能产业时代，还有一个重要的评价维度，那就是城市数字化、智能化发展的水平。

所幸，在实现智能化发展这条道路上，重庆这座城市并没有花太长时间。当下，重庆正迎来物联网、人工智能、大数据、云计算等新一轮技术浪潮，创新科技已成为引导重庆变革的动力。借助科技，重庆的资源已被有效地整合利用，从而迈入了城市的精细化和智能化管理进程。

当你伫立在嘉陵江边的重庆，铺开重庆近五年的时光卷轴，会读到一种由人、技术、城市组成的全新交融，这其实也是一张以智能化为生活添彩的答卷。

"人气之城"

判定城市价值的一个直观标准是人气。一座城市的人气足够旺，有源源不断的外来人口流入，就说明这个城市较为开放，且经济向好、产业优良，有足够多的就业发展机会。

"到西南去，攻占重庆。"这是几年前创业者们喊出来的口号。

理想的经济和营商环境，对各种生产要素具备更强的吸引力。2020年6月，中央广播电视总台发布的《2019中国城市营商环境报告》显示，在国内36个城市营商环境综合评价排名中，重庆位列第五。[1]

在国内主流媒体构建的重庆城市形象中，工业建设之城、新兴腾飞之城和希望之城成为重庆的基本画像；而在国际视野中，重庆又是一座怎样的城市？在众多海外媒体的笔下，重庆这座城市拥有众多不同的解读视角和琳琅满目的城市标签。

比如，韩国《今日周刊》用"长江经济带的重点城市"介绍重庆，而行业媒体如全球希腊航运新闻网、英国铁路货运公司网站、挪威奥斯陆交通和通讯网均有大篇幅报道介绍重庆与国际海陆贸易走廊和中欧铁路的连接。

在德国新闻社、德国之声等德国媒体以及韩国《首尔经济》以及英国《汽车物流》杂志的笔下，因为重庆是"连接中国和欧洲的重要铁路枢纽"，报道中更看重中欧班列的比较优势。

除了特殊的地理位置与交通优势，重庆城还因为独特的创新建筑而著称于世。除了识别性极强的洪崖洞等特色建筑，来福士广场等重庆建筑吸引了大量国际媒体的关注。美国在线、新加坡《联合早报》、沙特新闻网、美国《建筑师报》、葡萄牙《金融日报》、越南《劳动报》等媒体均有相关新闻报道。

而随着2018年，中国国际智能产业博览会在重庆举办并永久落户，吸引了国际媒体的大量报道，重庆收获了"科技""创新""智慧""智能"等新标签。美联社报道称，重庆成为媒体焦点不在于它的美丽风景，而是因为它的智能化优势。[2]

1　肖福燕，重庆日报，《重庆营商环境为何能进入全国前五强》，2020年8月21日。

2　刘昊、尹佳、秦昕婕，《"一带一路"倡议背景下重庆在国际媒体中的形象研究》，2021年7月20日。

● 眺望重庆城区
Overlook the downtown of Chongqing

直辖 25 年来，重庆更是以突飞猛进的速度，时刻更新着整个世界对于重庆的认知：

2010 年，重庆与北京、天津、上海、广州一起，被确定为国家五大中心城市；

2015 年，中新互联互通项目花落重庆；

2018 年，重庆开始举办智博会，吹响了经济转型升级的号角；

2021 年，重庆与上海、北京、广州、天津一道率先开展国际消费中心城市培育建设。

最近这五年，重庆引进中国 100 强民企、世界 500 强外企的速度越来越快，已经成为内陆现代服务业发展先行区：2018 年，腾讯西南区域总部、复星集团西南总部等落户重庆；2019 年，圆通速递西部总部、紫光 DRAM 业务总部落户重庆，字节跳动入驻重庆市渝中区；2020 年，去哪儿网第二总部、人人视频总部、启迪数据云集团全国总部等落户重庆；2021 年，世界 500 强企业正威国际集团把第二总部设在重庆。

看好重庆的创业环境以及消费升级的潜力，这也是各大知名企业集体落户重庆的根本原因。

城市由人组成。一个充满活力与连接、多元包容、健康生态的城市，更是激发交流、创新与合作的摇篮。

在人口指标上，重庆再现增长。2021 年 5 月 13 日，重庆市政府新闻办举行重庆市第七次全国人口普查新闻发布会，公布重庆市常住人口为 3205.42 万人，与 2010 年相比增加 320.80 万人，增长了11.12%，增长率是全国人口增幅的两倍多；10 年来跨省流入人口增加 115.16 万人，对人才吸引力增强，流入人口规模进一步加大。[1]

人口、劳动力、人才作为市场要素之一，不仅带动产业结构调整，也是城市发展核心竞争力的重要体现。而超越数万的常住人口增量，对重庆这样的成长型城市而言，或许还只是一个开始。

"技术之城"

数年之前，重庆还不是互联网行业的沃土。

近五年来，智慧之城成为重庆的又一"新名片"。

智慧城市发展理念是新技术变革与城市发展新挑战的共同产物，其本质是用技术的手段赋能城市，重塑城市的发展模式。这决定了智慧城市运营商在技术、数据、场景、资金上具备足够的势能。

---

1 光明网，《9 组数据，读懂 10 年间，重庆人口变化！》，2021 年 5 月 13 日。

解码智能时代
从中国国际智能产业博览会瞭望全球智能产业（2018—2022）

"城，所以盛民也。"随着城市的发展变迁，现代城市在城市理念、城市功能、城市辐射力等方面都发生了翻天覆地的变化。而这些年，重庆在"全局"中思考，在"大局"中行动，依靠智能化技术给城市带来了更多改变，缔造出更加美好的城市生活空间。

过去 5 年，智能化真实地改变了城市生活，也改变了"智慧城市"这个原本已经巨大的市场。如果你在城市生活，又足够关心身边这些基础设施以及市政新闻的话，大概不难发现，早在 2017 年人工智能就已经融入了中国的城市生活，比如便利店刷脸支付。

在 2021 智博会现场，百度交通大脑和永川智慧交通案例成果受到行业内的广泛关注。

以交通信号自适应为例，百度在永川重点区域 29 个路口进行了交通信号优化升级，复用已建设的外场感知设备，实现了城市道路交通状态全息精准感知和分析研判，打造出西部地区第一的精细化调控信号配时服务。目前已经实现 26 个路口的实时自适应控制，1 条干线的自适应动态滤波控制。试点道路平均车速提升约 5 公里 / 小时，停车次数减少约 20%。

永川实现了西部地区首个百度地图 app 发布路口信控信息服务。目前完成的 9 个信控优化路口的红绿灯状态、倒计时信息可实时推送到百度地图，引导市民及时调整车速，助力缓解城市交通拥堵，实现道路与车辆智能交互。此外，在智能公交服务上，百度地图 app 设置独立的实时公交入口，接入永川区 28 条公交线路，实时提供包括公交站点、线路、到站时间在内的信息服务，方便居民出行，让群众直接感受智慧交通建设成果。[1]

关于自动驾驶，在全球范围内存在各种不同技术实现的流派，其中既有"智能车"流派，也有"智能路"流派，显然，百度与永川合

1 解小溪，重庆日报，《百度亮相重庆 2021 智博会 重庆永川智慧交通项目成果引人瞩目》，2021 年 8 月 26 日。

作探索的智慧交通方案，已经成为汽车厂商之外，对于智慧交通最为前沿的探索。

重庆，这座智慧城市的应用变革早已无处不在。

"智慧之城"

放眼整个中国，早在五年之前，智慧城市战略几乎已经成为一种时代显学。

智慧城市的概念，最早可以追溯到 IBM 提出的"智慧地球"。2008 年 11 月，IBM 在美国发布的《智慧地球：下一代领导人议程》主题报告中提出了"智慧地球"的概念，意思是把新一代信息技术充分运用到各行各业，提升整个社会的数字化水平。

当时，这一概念得到了广泛的认同，而 IBM 也非常注重全球城市化速度最快的中国，特地将报告搬到中国又发布了一次，向中国市场推介"智慧地球"以及智慧城市的概念。

2012 年，住房和城乡建设部发布了《关于开展国家智慧城市试点工作的通知》，智慧城市开始在全国范围内陆续铺开，及至 2017 年底，中国已经有超过 500 个城市明确提出或正在建设智慧城市。

大小城市们面向未来的集体信仰和共同实践，让整个中国的智慧城市建设保持着高速发展。

到了 2021 年 3 月 11 日，国家"十四五"规划和 2035 年远景目标纲要中，已经明确提出："分级分类推进新型智慧城市建设，将物联网感知设施、通信系统等纳入公共基础设施统一规划建设，推进市政公用设施、建筑等物联网应用和智能化改造。"这代表着智慧城市已经从局部试点阶段正式进入了全面统筹推进的阶段。

那么，从全球范围来对比，中国智慧城市的发展水平又到底如

● 2021 智博会期间举办的自动驾驶汽车挑战赛，近 30 名车主驾驶量产车参赛
Self-driving car challenge was held during the 2021 Smart China Expo, with nearly 30 owners attending the competition.

何？ 2021 年 12 月 2 日，德勤发布《有目标的城市未来：2030 年塑造城市未来的 12 种趋势》报告，报告通过全球范围的城市调研和观察，揭示了 2030 年城市生活的 12 种发展趋势。报告经对比分析后发现，中国在低碳智能出行、以人工智能实现城市运营自动化、利用人工智能实施警务预测等领域的渗透率，已趋于全球领先水平[1]。

而作为智能时代的坚定探索者，重庆的智慧城市建设，已经取得了诸多成绩。

几年前，让人"跑断腿"的政务服务，如今正在实现"秒报秒批一体化"。重庆市民通过城市云平台，让数据多跑路，百姓少跑路，

1　王恩博，中国新闻网，《专访德勤中国副主席：智慧城市，"聪明"就够了？》，2021 年 12 月 04 日。

不断刷新"不见面审批""全城通办"范围。另外，对公业务方面，为持续优化营商环境，企业开办审批时间压缩到几十秒以内；企业项目备案，从申请到批复只需要 6 秒。而这些，都是"政务上云"的功劳。气象预报、供暖、发电、自来水系统监控等，无不渗透着智能化能力。

重庆对科技的重视和笃信，为这座城市的智能化发展打下了坚实基础。而我们也可以观察到，数字化、智能化对于重庆的城市治理与经济发展起到了一系列显著的促进作用。

在重庆市发展规划中，智慧城市将不再仅仅是城市的锦上添花，而是城市治理能力与治理效率的核心组成部分。"一图全面感知、一键可知全局、一站创新创业、一屏智享生活"等目标已经非常清晰，越来越多的智慧应用，让重庆居民享受到前所未有的便捷体验。

全场景智慧，正成为未来城市通行证。实现城市整体智能化发展成为重庆的核心方向，各部门之间实现数据打通，互相给予智能判断与智能决策，走向真正的"城市大脑"。比如疫情防控期间就涌现出了健康码、密切接触者追踪、复工复产、线上服务等各式各样的应用场景，应用创新也在反向驱动云基础设施的更新和优化，要求更多业务及应用一开始就诞生在云端、沉淀在云端、生长在云端。

重庆，在蜿蜒流淌的进程里，同江水一起生生不息，滋润万物生长，灌溉文明勃发。这座山水之城，在一片盛放的数字化花海中，邂逅未来。

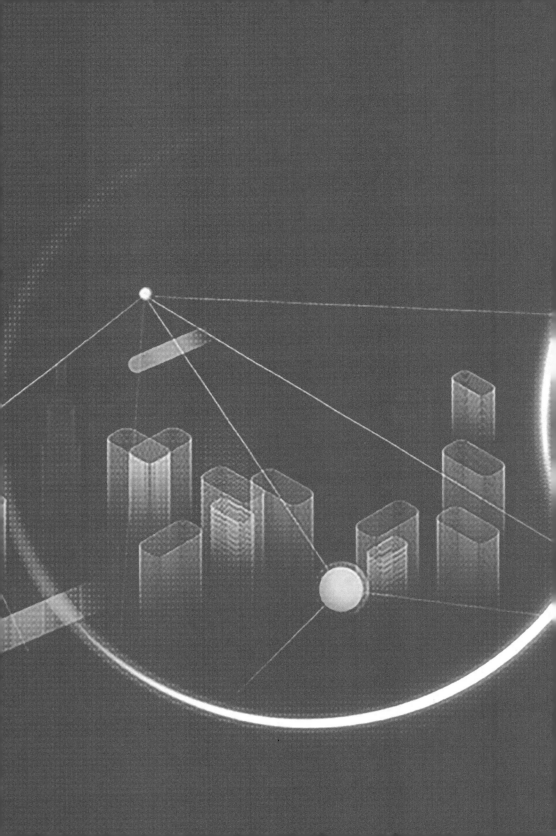

第四章

智慧治理：建设智慧城市的智能化中枢

　　一个个日趋完善的智慧系统，一个个精彩绝伦的智能场景，一系列以人为本的智慧应用，城市的智能化治理，仿佛成为一个拥有思考能力的生命，既从宏观层面提升城市的整体运行效率，又从微观层面赋予市民细腻的幸福体验。

　　智慧城市的治理方式，在人工智能的加持下，正在从"治理"走向"智理"，而不同城市在智能化治理方面呈现出的不同侧重，又形成了智慧城市之间不同的城市性格。

第一节
不同的时代，
全球智慧城市的不同定义与思考

城市起源于哪里？标准不同，尚有争议，但人们普遍认为8000多年前的恰塔尔休于是人类历史上最古老的城市之一。

恰塔尔休于城里有1000多座土砖砌的房屋，人口超过6000人。经过农业革命、工业革命和信息革命的发展，8000年后的城市，无论量级规模，还是社会功能，都已经完全不同。

据我国第七次人口普查数据显示，中国人口超过2000万的城市有四座，分别是重庆、北京、上海和成都，而重庆在其中位列第一，常住人口有3200万人之多。

今天的城市，人口和房屋与最初的城市相比，有"核弹级爆发式"的增长，城市形态也更加多样化，包含小城市、大城市、国际大都市、城市群等，更为不同之处在于，城市的内涵、运转的物质结构、治理方式的不同。

城市的文明程度，早已发生了翻天覆地的变化。

有人说，城市是人类最伟大的发明。大概是因为城市激发和承载了人类的所有文明。而在全球范围内，关于城市未来的最新共识是，人类城市的文明走到了基于大数据、云计算、智能化的智慧城市阶段。

不同时代的"智慧"城市

假如有一位土耳其恰塔尔休于的居民,从 8000 多年前穿越到后来的各个时代,他对智慧城市的理解一定会大不相同。

因为,每个时代对城市的"智慧"有着不同的注解。

在 21 世纪到来之前,有很多事物或许会被古老的恰塔尔休于居民认为是城市的"智慧"。

比如穿越到公元前的罗马,下水道被发明并成功运转、城市告别脏乱不堪的境地,他一定会认为这就是智慧城市,它更文明,也让人类远离更多疾病传染的可能。

如果穿越到 150 多年前的伦敦,当他第一次踏上地铁并被它快捷地送往某个目的地的时候,他心里也会感叹:智慧的城市应该如此。

更别说 100 年前的纽约,这个城市的电网开始连接千家万户的电灯与收音机。穿越过来的恰塔尔休于居民,看着黑夜中千家万户的灯光,听着收音机里丰富多彩的节目,或许会短暂地感到无所适从,但适应之后一定会感慨:这就是一个智慧城市该有的样子。

最让古老的恰塔尔休于居民无法理解的,则是 21 世纪之后的现代城市。这个时候的城市,电网之上又叠加了互联网;互联网自身也在发展,发展出了移动互联网。互联网、移动互联网带来的产业数字化和数字产业化,开始让城市里的物"说话",而后让物与物、人与物可以"对话"。同时,越来越多的物,不仅与人对话,还帮助人更好地做决策,当家作主。

这便是新的智慧城市时代,城市如同拥有了智慧大脑。

关于智慧城市的概念,最早出现在 2007 年 10 月欧盟发布的《欧洲中等城市排名报告》,其中提出了智慧城市 6 个方面的内容框架,分别是智慧经济、智慧流动、智慧环境、智慧公众、智慧居住和智慧管理。

而 2008 年 11 月份，IBM 公司正式提出智慧地球的概念，并于 2009 年在全球 50 多个国家推行"智慧地球"概念和"智慧城市"解决方案，智慧城市一词，从此开始确立，并逐渐被世界认知和接受。

通过维也纳理工大学、卢布尔雅那大学和代尔夫特理工大学于 2007 年联合发布的《欧洲中等规模智慧城市的排名》，可以看出，最初的智慧城市是通过划分功能模块的方式来定义的，模块有经济、环境、居住等。这代表了一种定义的方式，只是随着经济的发展、技术的演进，越来越多的社会功能开始被"智慧化"，定义智慧城市的模块，即维度划分，也就越来越多，越来越细，如智慧政务、智慧教育、智慧安全、智慧制造等。

智慧城市给市民生活带来哪些变化呢？

虽然关于智慧城市的不同定义，各种框架角度不一，各种标准考量有别，但从智慧城市的根本上有三个角度可供观察，那就是智慧城市可以在互联、高效和便利三个方面为城市带来多大的提升。

首先，智慧城市应该是互联的城市，细分为整体性的全面感知、跨领域的立体互联和执行层的共享协同，一座城市的基础设施，基于新的通信技术，互联程度越高，其智慧开发的潜力也就越强。

其次，智慧城市应该是高效的城市，细分为节能低碳、灵敏便捷和整合集群三个方面，城市存在的意义，就是以人类的大规模聚居，实现社会资源的高效利用和城市生活的高度便利，这始终是衡量城市发展水平的关键。

最后，智慧城市应该是便利的城市，又细分为无线泛在、就近一体和个性互动三个方面，通过大数据与通信技术的维度，为智能生活创造无限的可能。

每个时代对城市智慧的理解不同，同一个时代对智慧城市的定义也有不同的发展阶段和侧重点。但有一点可以非常明确，新的智慧城市建设正在如火如荼展开，人类的城市又走进了令古人惊叹的变革之中。

甚至可以说，这次的变革是历史上范围最广、变化最大的一次变革。

全球城市的智慧思考与实践

自 2009 年 IBM 提出"智慧城市"，各国开始逐步布局智慧城市的建设，投入金额逐年升高。

根据 2020 年 7 月，管理咨询公司德勤发布的《超级智能城市2.0，人工智能引领新风向》报告，对智慧城市的投资金额，亚洲的复合增长率为 12.2%，北美的复合增长率为 8.6%，欧洲的复合增长率为 9.65%，预计将在未来几年内达到高峰。

国别不同，国情不同，发展路径和实现手段，自然不同。

日本在 2009 年制定了《i-Japan2015 战略》，日本的港口城市横滨对此进行了积极的实践。日本是个资源相对匮乏、同时频繁遭受自然灾害侵袭的国家，基于如此基本国情，横滨的智慧城市建设重点为"能源节约"。

具体措施上，包括引入可再生能源与电动汽车，对家庭、建筑物和小区实施智慧能源管理，为市民设置家庭能源管理系统，呼吁参与实施大规模节能行动实验……

荷兰于 2008 年推出 Amsterdam Smart City 智能城市计划。阿姆斯特丹是荷兰最大的城市，世界上最重要的港口之一，智慧城市建设方向为"改善环境、节约能源、建设可持续公共空间"。

阿姆斯特丹在推进智慧城市建设中，最值得一提的执行思路是"开源与共享"，即专门建设了一个超级数据库，整合了所有城市区域超过 12 万个数据集，并对该数据库进行开源，让每个市民或企业家都能基于该数据的使用参与城市的智慧化建设，享受城市的智慧化

● 2021 智博会新加坡展台
Singapore booth at the 2021 Smart China Expo

成果。

　　数据开源，是一个激发大众进行城市智慧化建设的值得借鉴的举措。

　　巴西的狂欢节之都——里约热内卢，建设智慧城市的核心诉求是保障公共安全。为此，里约热内卢在市内安置了大量的监控探头和无数的传感器设备，从而清晰地掌控整个城市的情况，对突发事件及时作出反应和部署，保障公共安全。

　　美国最大城市及最大商港——纽约，同样是智慧城市建设的先锋。纽约在智慧交通系统领域的创新实践值得关注，打造的交通信息服务系统可以及时跟踪、监测全市所有交通状态并预估其动态变化，车主可以根据系统选择最佳行驶路线，相关部门也可以根据后台的路况信息进行交通引导与疏通。

新加坡于 2006 年启动了"智慧国 2015"计划，又在 2014 年公布了"智慧国家 2025"的 10 年计划。新加坡正将整个城市转换成一个虚拟模型，即在网络世界中呈现一个数字化的新加坡，不同的城市管理部门可以根据该模型实现对城市动态的监测和协同管理。

新加坡的数字化尝试，正是基于数字孪生技术的应用。这项技术已越来越多地被用在城市管理和智慧城市建设上，这将为城市治理带来高效、精准的优势。

不同城市对智慧城市的不同理解，不同城市对智慧城市各有侧重的实践，共同搭建了这个时代的智慧城市全景，各个城市可互相参照、互为借鉴，继续完善人类的智慧城市建设。

这个智慧城市的全景中更有不可忽视的一抹亮色——始于 2009 年的中国的智慧化城市建设。凭借着在移动互联网时代积累的大量数据、成熟的云技术能力、扎实的类似 5G 的基础设施建设，中国正在与全球智慧城市建设共享先进方案。

第二节
北上深杭，
中国城市治理中的智能化创新应用

这是毋庸置疑的事实：穿越历史云烟，回望人类城市的发展进程，那些代表人类最高智慧的城市，一定有不少来自中国。

比如，7 世纪、8 世纪的中国长安（今西安），被称为"世界性的都市"，常住居民逾 60 万之多，算上客商、僧侣等，人口总数极有可能超越 100 万，城市面积 80 多平方公里，与 300 多个国家和地区有联系，是当时最繁华的城市。

比如，11 世纪、12 世纪的中国汴州（今开封），面积约 34 平方公里，但人口总数却达到了 140 万左右，已经是当时世界上人口数量最多的城市，城内有 8 万多名各类工匠和 2 万多家商店，每日车水马龙，工商贸易极为繁荣。

曾经的标杆城市，评价的维度是人口、面积和商业，而随着智能科技的进步，智慧化则成了评价城市的另一个重要维度。

同样，如果要在 21 世纪林立的都市中，评选出最具智慧化特征的城市，仍然有不少来自中国。

解码智能时代
从中国国际智能产业博览会瞭望全球智能产业（2018—2022）

中国城市的智慧化进程

建设智慧城市的意义和路径，或有不同的认知和实践，但对建设智慧城市这一方向，对智慧城市前景的认可，在当下的中国是有广泛共识的。

中国的城市已大面积展开智慧城市的建设。

根据德勤2021年发布的报告显示，自2018年起，全球已启动或在建的智慧城市达1000多个，其中中国就占了将近一半的数量，约有500个城市正在推进智慧城市建设的试点。

如此蓬勃的智慧城市建设态势，离不开各级政府的重视和推动。

从2016年到2021年，国家陆续出台《"十三五"国家信息化规划》《"物联网与智慧城市关键技术及示范"重点专项2019年度项目申报指南》《2021年新型城镇化和城乡融合发展重点任务》《关于推广第三批国家新型城镇化综合试点等地区经验的通知》等政策，明确了智慧城市作为我国城镇化发展和实现城市可持续化发展方案的战略地位，以及"推进智慧城市建设"的任务，刺激了各地对智慧城市的建设需求，对促进智慧城市和相关行业的发展具有极大的积极作用。

从最初的概念提出，城市试点，推广执行，到具备成熟思路和成功案例，中国用了13年时间。

2012年底，中国住建部推出首批国家智慧城市试点，包含北京市东城区、北京市朝阳区、上海市浦东新区、杭州市上城区、深圳市坪山新区、重庆市南岸区、重庆市两江新区等共90个试点城市（区、镇）。

2014年，国家发展改革委、工业和信息化部、科学技术部、公安部、财政部、国土资源部、住房和城乡建设部、交通运输部八部委印发《关于促进智慧城市健康发展的指导意见》，提出到2020年建设一批特色鲜明的智慧城市。

同年，26 个部门联合成立了促进智慧城市健康发展部际协调工作组，这是在机构设置层面上确保智慧城市发展高效进行、顺利开展。

2016 年，国务院印发《"十三五"国家信息化规划的通知》，正式提出新型智慧城市建设的行动，明确牵头单位为国家发展改革委和中央网络安全和信息化领导小组，从而正式确定了新型智慧城市的发展方向，将建设新型智慧城市确认为国家工程，推动智慧城市建设进入快车道。

而围绕智慧城市建设，中国已经发展出完整的产业链，并有对应的代表性企业。

按照艾瑞研究院的分类，智慧城市产业图谱可以分为 5 层。

第一层是智慧城市顶层设计层，包含华为、太极股份、神州数码、浪潮、中电科等，为智慧城市建设提供整体解决方案。

第二层是感知与通信层，包含中国移动、中国电信、中国联通、中兴、大唐电信等企业，换言之，这些企业主要进行数据的收集与传输。

第三层是平台与基础设施层，包含阿里云、平安科技、新华三、中科曙光、太极股份等。

第四层是城市计算层，包含京东数科、阿里云、腾讯云、百度云、科大讯飞、平安科技等，简单来说，这是提供算力和算法层。

第五层是细分场景应用层，也是居民最容易感知的一层，比如建设智慧交通就离不开海康威视的监控器材，类似的应用层企业还有千方科技、佳都科技、新中安、易华录等。

此外，还有一些专攻城市安全的企业如 360、深信服、亚信、安恒信息、瀚私等贯穿产业上下游。

目前，国内智慧城市主要的建设参与者，依然以科技大企业为主，其中京东数科、阿里云、腾讯云、华为、平安国际智慧城市、浪潮等，因为它们既有领先的解决方案，又有强大的技术实力，因此可

● 2021 智博会，汇集了国内外众多智慧城市产业链相关企业

At the 2021 Smart China Expo, many domestic and foreign companies related to the smart city industry chain were brought together.

以充分覆盖智慧城市产业图谱中的不同层面。

　　智慧城市建设是一项浩大的工程，需要科研、技术的应用，更需要政府、商业公司、科研机构和社会公众的广泛参与。

　　当前，国内智慧城市的建设，遵循了市场导向：以政府投入为主体，辅以与实力强大的商业公司合作。战略规划与顶层设计由商业公司、科研机构和智库进行，以确保方向的正确与实践的成效；应用领域开发则是政府与商业公司紧密配合。

　　总之，社会各界的广泛参与，正推动着中国智慧城市的蓬勃发展。

北上深杭的智能化创新应用

　　中国每个城市的基础条件、资源禀赋不同，发展智慧城市的路径和侧重点也有差异。正因为各城市之间的差异，在中国能找到多种智慧城市发展样本。通过不同城市样本的差异化探索，中国极有可能为全球智慧城市建设提供更接近理想目标的参照。

北京致力于建设全球新型智慧城市的标杆。

一方面，北京持续加强城市的"感知能力"，推进智慧杆塔等感知底座组网建设，实现多种设备和传感器"一杆多感"综合承载；积极推进千兆宽带接入网络建设，加快基于互联网协议第6版（International Protocal Version 6，IPv6）的下一代互联网部署，为物联网、数据传输、自动驾驶等扫清数据、信号传输上的障碍。

另一方面，为了使城市治理更有效，北京正在构建"一网统管"的应用体系，构建以城市人口精准管理、经济活动监测、城市运行感知、城市管理综合执法等为核心的城市运行管理与决策支撑体系。基于这样的体系，决策将会依据更全面、更及时的信息，城市的治理将会更高效、更精准，实现数据驱动的智能化。

不同的城市，相似的智慧城市运行理念，北京有"一网统管"，而上海有"一网通办"。

"一网通办"是上海的政务服务品牌，据2021年10月统计数据显示，上海接入"一网通办"平台的事项已超过3400项，日均办件量16万件[1]。随着越来越多的管理部门将其公共服务事项接入"一网通办"，这个数据量还会增长，背后则是更加便利、舒适的城市生活，更加公开、高效的营商环境。

上海的智能化治理中，有很多人性化的设计。如智慧电梯平台，借助远程监测自动识别将上海电梯困人应急救援平均时间缩短至12分钟[2]。在上海长宁区，独居老人家中安装的智慧水表，若12小时内读数低于0.01立方米，系统会自动通知街道办事处，以便其第一时

从中国国际智能产业博览会瞭望全球智能产业（2018—2022）

解码智能时代

---

1 周頔，澎湃新闻，《上海："一网通办"日均办件量16万件，持续深化数据治理应用》，2021年11月19日。

2 刘浩，《中国消费者报》，《2021年"上海标准"发布：电梯应急救援时间缩短至12分钟》，2021年10月25日。

● 上海市的智能化治理，融入了很多人性化的设计
Intelligent governance in Shanghai incorporates many humanized designs.

间派人上门探视情况。[1] 利用这些小的创新应用，上海能够更智慧化地解决掉一个个细微的问题，一点点构筑出更加安全、体贴的城市环境。

现在，深圳市政府携手华为建设的城市智能运营中心（Intelligent Operations Center，IOC），是"数字政府"的"大脑"。这是一个能看、能用、能思考的智慧城市中枢，打通了 42 个系统，100 多类数据，28 万多路视频，构建起深圳市的"1+12+N"的一体化指挥体系，形成了坚实有力的政府数据"底座"。[2]

深圳市智慧城市的创新探索，带着鲜明的政企合力印记，每个企业都能在智慧深圳的建设中找到属于自己的角色。在深圳，腾讯参与打造的"粤省事"和平安国际智慧城市承建的"i 深圳"都在实现政

---

1 李舒，上游新闻，《12 小时内不足 0.01 立方米就报警，智慧水表如何读出独居老人的生活异常？》，2020 年 12 月 19 日。
2 王新根，深圳晚报，《解读一座创新之城的"智慧密码"华为助力深圳智慧城市建设》，2020 年 5 月 21 日。

务服务由"线下办"到"线上办"、"网上办"到"掌上办"。华为与深圳市水务局、中国水利水电科学研究院联合搭建的多维水务模型体系，将实现晴天维持河湖生态健康、中小雨减少初雨污染、大雨减轻内涝灾害、暴雨确保城市防洪安全等预期成效；与深圳地铁打造的全球首个线网级工程数字化管理中心（CDMC），创新实现全城400多个地铁建设工地的集中管控，全面提升城轨交通建设的信息化管理水平。[1]

早在 2016 年 4 月，杭州就开启了一项名为"城市大脑"的人工智能探索，只不过最初，还仅仅是以交通领域为突破，利用大数据改善城市交通。也正是从那时候开始，杭州迈出了从"治堵"向"治城"跨越的步伐，陆续建成了覆盖公共交通、城市管理、卫生健康、基层治理等 11 个领域的智能平台，囊括 38 个应用场景、366 个办事事项，能找车位、查天气，也提供健康码、民意直通车、12345 民生平台[2]，让杭州成为一座数字系统治理之城。

2022 年，杭州城市大脑迎来了 2.0 版本，以社区为切入点，首次聚焦"住""行""老""小"等领域，推进住房、交通、未来社区中"一老一少"等重大项目建设。

沉淀最久的智慧交通场景，在城市大脑 2.0 中，将建设"智慧规划、智慧交管、智慧公交、智慧地铁、智慧物流、智慧停车、智慧慢行"七个子场景，协助打造畅通、安全、便捷、高效、绿色的城市交通。

---

1 周雨萌，深圳特区报，《2020 年即将收官，深圳智慧城市建设交出亮丽答卷》，2020 年 12 月 28 日。
2 黄平，经济日报，《给城市装上"大脑"——杭州市智慧城市建设调查》，2021 年 3 月 23 日。

第三节
解读智慧重庆，城市治理智能化的突飞猛进

关注 2019 智博会的重庆市民，一定还记得两个词，一个是"智造重镇"，另一个是"智慧名城"。

2019 年 8 月 27 日，在 2019 智博会的开幕式上重庆提出：集中力量建设"智造重镇""智慧名城"，用实干实绩共创智能时代、共享智能成果。

这两个词，既是打造智慧城市的重庆标签，也是迈向智能时代的重庆愿景，既着眼于重庆的经济发展，也考虑到重庆的生活体验，与智博会"为经济赋能、为生活添彩"的主题遥相呼应。

将"智造重镇"与"智慧名城"作为重庆智慧城市建设的远景目标，背后是重庆智能化能力和治理的突飞猛进，也是重庆在智能时代的底气和决心。

智慧重庆的"任督二脉"

翻阅历史，可以说重庆的智慧化城市建设发轫于 2000 年，而提速于 2018 年。

2000 年，重庆被国家建设部（现住房和城乡建设部）批准为"数字化城市"示范市，彼时的数字重庆建设，更多是对物理层面的城市

进行数字化的记录，也就是将城市上网。

比如，建成的数据库包含基础地理信息数据库、地下空间数据库、地名地址数据库、综合管网数据库等，建成的软件平台包含基础地理信息平台、政务地理信息平台和社会服务地理信息平台。

"地理"是数字化重庆建设的关键词，下一步的关键词则是"物联网"。

在物联网方面，重庆市先后编制和颁布了《重庆市"十二五"物联网产业发展规划》《重庆市人民政府关于加快推进物联网发展的意见》。

从"地理"到"物联网"的转变，透露了重庆数字化建设的重大进展，即在 21 世纪前十年，重庆完成了对城市地理与空间的数据化，并逐步将数据化的范围扩大到城市治理、城市生活的方方面面，于是"物联网"成了关键词，要将万物联网并沉淀数据，实现更多维度的智慧化探索。

可以说，从"地理"到"物联网"，是实现"平面重庆的数字化"到"立体重庆数字化"的一个过程，有赖于传感设备的发展，数据维度和数据量，有了质的变化，云计算就显得更加重要，因此，重庆市又开始建设两江国际云技术中心，并且成立了重庆市云计算产业协会。

时间走到 2012 年底，住房和城乡建设部公布首批国家智慧城市试点，重庆市两江新区、南岸区均被列入。智慧城市概念，逐步在全国各地普及。

2018 年，在重庆发生了两件大事，开启了智慧重庆建设的新阶段。其一是 2018 智博会在重庆成功举办，这对重庆的智慧城市建设影响深远，体现在汇聚财智、汇聚企业、汇聚产业、汇聚共识等多个层面。另一件则是重庆市大数据应用发展管理局挂牌。

作为中国首批设立的 8 个大数据局之一，重庆市大数据应用发展管理局不仅肩负大数据管理相关职责，也肩负智慧城市建设的职责。该机构的设立，其意义不仅仅在于智慧城市建设有了专门的负责机构，更在于与后来重庆推出的"云长制"一道，打通了智慧重庆建设的"任督二脉"。

"云长制"，即 2019 年重庆市出台的《重庆市全面推行"云长制"实施方案》，市政府主要领导任"总云长"，在政法、交通等 6 个系统设"系统云长"，各区县政府、市级各部门主要负责人为各单位"云长"的架构体系。

智慧城市建设一方面有赖于基础设施、硬件水平等硬实力，另一方面有赖于应用思路、数据等软实力。数据至为关键，被称为智能时代的动力与血液，重要性犹如电力之于电气时代。前文回顾的重庆智慧城市建设的历史，本质上是一场数据的收集、计算和应用的历史，其不同在于数据的收集能力、计算能力、传输速度、应用场景等。

智慧城市建设的基础是数据共享，数据共享则分为收集与共享两个步骤。简言之，在数据层面，"云长制"确保了各部门、各区县的数据上云，并满足一定的标准和条件，这是解决数据的有无和质量问题。而大数据局的牵头，则能很好地统筹数据的应用，解决数据的互联互通与数据的共享问题。

数据的收集上云和共享应用，正好对应"云长制"的机制功能和大数据局的机构职责。据报道，2021 年，重庆市实现政务数据共享 4055 类、汇聚 3309 类、开放 1310 类，累计调用数据 142 亿条，而每一条数据的调用和应用[1]，都推动着重庆治理逻辑和生活方式的智慧化。

1　崔力，重庆日报，《聚焦解决"痛点""堵点""难点"打通底层数据 重庆"智慧名城"建设加码》，2022 年 2 月 24 日。

智慧名城的建设体系

"云长制"和大数据局从机制机构层面，解决了数据相关的痛点、堵点，智慧城市建设的"血液"从此能够畅通无阻，"动力"能够尽情释放。

对智慧城市建设中数据这一关键要素的赋能，为重庆近些年智能城市治理突飞猛进奠定了一大基础。数据是智慧城市建设的一个非常关键的点，而整个城市的智慧化发展，还有赖于整体规划、全面发力。

2019 年 4 月，重庆通过《重庆市新型智慧城市建设方案（2019—2022 年）》，提出以"135"总体架构推进重庆新型智慧城市建设，即建设由数字重庆云平台、城市大数据资源中心和智慧城市综合服务平台构成的 1 个城市智能中枢，夯实新一代信息基础设施体系、标准评估体系和网络安全体系 3 大支撑体系，发展民生服务、城市治理、政府管理、产业融合、生态宜居 5 类智能化创新应用。[1]

延续上述建设方案的思路，重庆正在加快构建"8611"一体化场景建设体系，"智慧名城"的建设路径更加细化，建设步伐正在加快。

"8"指 8 大基础能力，指加快建设全国一体化算力网络成渝国家枢纽节点、"三千兆"城市、AIoT 基础设施、城市信息模型平台、"山城链"，升级城市大数据资源中心、数字重庆云平台，推广中新国际数据通道，提升数字、网络、算力等场景建设基础能力。

区别于"云长制"和设立大数据局对数据的机制体制上的赋能，8 大基础能力，实际是从技术层面赋能数据，以提升对数据的传输速度、储存和计算能力、丰富数据收集维度、共享场景等。

"6"指 6 大支撑体系，通过推动"数据目录、数据标准、逻辑架构、系统接口、业务流程、能力组件"六统一，持续提升数据支撑能

1　重庆日报，《市政府召开第 45 次常务会议》，2019 年 04 月 10 日。

解码智能时代
从中国国际智能产业博览会瞭望全球智能产业（2018—2022）

力、系统融合能力、业务协同能力，夯实场景融跨共性支撑。

第一个"1"指 10 个以上融跨平台，渝康码即"融跨平台"之一，实现了跨层级、跨地域、跨系统、跨部门、跨业务的协同管理和服务。

"重庆健康出行一码通"是在疫情防控的特殊需求下，重庆智慧城市建设结合大数据与云计算而诞生的典型创新，能为市民提供防疫相关信息的查询功能，如核酸检测情况、疫苗接种情况、行程情况等。其功能呈现有赖于两个方面，一方面是卫生健康、公安、工信、交通运输等部门的数据收集，另一方面则是对这些数据进行了共享和应用。

除了"重庆健康出行一码通"，"融跨平台"还包括为大众提供政务服务的"渝快办"，据 2021 年下半年重庆市的公开数据，可通过"渝快办"办理的事项已达 1875 项，用户突破 2100 万人。此外，"渝快政"在 38 个单位试点应用，省级政府网上政务服务能力评估排名进入前 10 位；"渝快融"用户超 26 万，为中小微企业融资 371 亿元。智慧城市的能力，已从服务大众，扩大到政务群体和中小微企业。

最后一个"1"指 100 个典型应用，即以应用为导向，打造"住业游乐购"全场景集。"'住'方面，重点打'未来社区'，包括智慧医疗、智慧养老院、智慧消防等。'业'方面，重点打造智慧教育、智慧就业等场景。'游'方面，重点打造智慧交通、智慧景区等场景。'乐'方面，重点打造数字娱乐体验、智慧体育等场景。'购'方面，重点打造智慧商圈、智慧金融、智慧物流等场景。"[1]

如果说类似"渝快办"这类融跨平台的增多正在让重庆的城市治理更加智慧化，那么围绕"住业游乐购"相关智慧应用的普及，则是让城市生活添上了智慧的色彩。

1　向菊梅，重庆日报，《重庆智慧城市建设水平走在全国前列》，2018 年 8 月 18 日。

● 迪马工业研发的智能消防机器人备受关注
Smart firefighting robots developed by Dima Industries attract attention.

新能源车若出现电池故障、电量不足，后方的动力电池风险实时监控与预警大数据平台可实时提醒；购买景点门票，预定住宿，"掌"上全搞定，戴上 VR 眼镜还可以穿越时空、游览全重庆；走进商场，点击触摸屏，所有门店介绍一目了然，还有机器人带你前往相应门店……

社区正越来越智慧。有些小区将人脸识别技术与小区门禁功能结合了起来，刷一下脸，大门立马打开。门禁系统还与公安机关的系统进行关联，实现对进出小区的人流、车流的实时监管、实时预警；一些小区通过安装智能监控系统，大大减少了高空抛物事件的发生；一些小区通过引入大健康物联网系统，实现了对老人、残疾人、小孩等人群的全天候健康管理和精准服务……

智能化的因子，正在智慧名城建设的整体规划下，随着数据、平台、应用……渗入重庆的每一个细胞，一个全新的、智慧的城市，正在加速到来。

第四节
智能化城市治理，未来的发展空间

在新型智慧城市建设发展要求的助推下，全国各城市治理水平取得显著提升。

深圳市的智慧城市建设，以"一图全面感知、一号走遍深圳、一键可知全局、一体运行联动、一站创新创业、一屏智享生活"六个一总体建设为目标[1]，数据的有效连接与智能交互打破了不同部门、领域之间的壁垒，破解了治理碎片化难题。

上海市加快智慧城市建设的文件中，细化了加快推进城市运行"一网统管"的具体要求，一体化建设城市运行体系，紧扣"一屏观天下、一网管全城"目标，加快形成跨部门、跨层级、跨区域的协同运行体系。

在重庆市的规划中，到 2025 年新型智慧城市运行管理中心的全面建成，也将推行城市运行"一网统管"、政务服务"一网通办"、应急管理"一网调度"、基层事务"一网治理"，全面提高数字治理水平[2]。

借助前沿技术、先进设备等智能化手段，智能感知、智慧识别、自动预警等技术被广泛应用，正成为城市治理精细化品质化的有力支撑。

1　周雨萌，深圳特区报，《深圳"六个一"理念 引领智慧城市建设》，2020 年 10 月 14 日。
2　向菊梅，重庆日报，《重庆将打造全国一流的数据集聚洼地和利用高地》，2022 年 1 月 5 日。

智能化城市治理，已经成为满足人民日益增长的美好生活需要的重要一环。

城市从"治理"到"智理"的演变

城市规模的不断扩大所带来的挑战，使城市治理复杂程度日益提升。

社区大型化、居民原子化、诉求多元化和可持续性发展等问题，成为城市治理的全新挑战。

本质上，城市是人和资源在有限空间的高度聚集与协同，随着人类跨入数据时代以及智慧城市的建立，城市在治理过程中的侧重点也在发生着某些转变。

● 重庆"智慧河长"遥感监测项目，数字赋能河道治理
Chongqing's "Smart River Manager" remote sensing monitoring project, digital technology empowers river governance.

如果说努力为人们创造宜业、宜居、宜乐、宜游的良好环境是城市治理 1.0，那么，关注人们的安全感、获得感、幸福感便是城市治理的 2.0。城市治理从"物的建设"上升到"人的服务"，引发了城市从"治理"到"智理"的演变。

智能时代的大数据与云计算，为城市治理开启了新篇章，于是我们看到，智能化技术开始为城市治理添上"智慧之翼"。

智慧治理的理念在于依托智能化技术，汇集众智对城市实施精细化治理。城市有了智能化技术这个"大脑"，治理也就更"聪明"：物联网、地理信息技术、网络通信技术、大数据、云计算和社会计算等关键技术在数据收集、数据传输和数据处理方面发挥着至关重要的作用。

智能技术用于城市治理的例子层出不穷。智能安防，为我们的安全保驾护航；智慧司法，助力维护社会的公平正义；智慧养老，以应对深度老龄化社会的即将到来……此外，城市的环境治理、公共卫生、交通出行等城市管理的方方面面，在智慧城市建设的统一调度、智能技术的强力加持下，也达到更低成本、更高效率和更好体验。

在具体操作中，治理主体能准确、便捷、及时地定位问题所在，实现靶向治理、精细治理。

过去，路灯坏了、井盖丢了、马路积水了，只能靠工作人员巡查或市民提供线索，很难在第一时间解决。现在，传感器等大数据设备成为城市的"大脑"和"眼耳手口"，只需通过连接智慧城市运营管理中心开展实时监管并及时发布监测到的异常情况，半小时内工作人员即可赶到现场处理。

在社区治理业务中，智慧治理推动城市运行的"一网统管"、政务服务"一网通办"，在提升政策宣传、民情沟通、便民服务效能方面颇有成效，真正让市民实现了"数据多跑路、群众少跑腿"。

城市治理未来的发展空间

按照智慧城市建设的愿景，城市精细治理能力培育的目标是在实现城市治理的决策、执行、监督和评估等方面都有证可循、有数可依的基础上实现整体治理。

所谓有数可依，流行着这样一种说法：公共数据"跑得快"，政府治理效率升。原因无他，政府部门掌握了 80% 的数据资源，开放政府数据有利于打通部门信息壁垒、释放数据价值，提升治理和公共服务水平。

在数字时代，数据已成为新的生产要素。

中国工程院院士倪光南在 2021 智博会上演讲时曾表示，当前，

● 2021 智博会上，中国工程院院士倪光南发表演讲

Ni Guangnan, academician of the Chinese Academy of Engineering, delivered a speech at the 2021 Smart China Expo.

解码智能时代
从中国国际智能产业博览会瞭望全球智能产业（2018—2022）

数据已经成为我国基础战略性资源，是重要的生产要素和生产力，数据的支撑对赋能新兴技术发展，促进产业数字化转型，提升社会治理能力和保障国家网络安全等方面具有重要意义。

智能化城市治理未来的发展空间在哪里？深耕数据治理尤为关键。

中国作为数字经济大国、数据治理大国，开展数据治理对驱动数字经济创新发展、全面提升政府治理效能，以及赋能公共服务和社会治理具有重大意义。根据市场研究机构 IDC 最新发布的《2021 年 V1 全球大数据支出指南》预计，全球大数据市场支出规模将在 2024 年达到约 2983 亿美元，五年预测期内（2020—2024）实现约 10.4% 的复合增长率（CAGR）；而中国大数据市场发展迅速，市场总量有望在 2024 年超过 200 亿美元，同时，中国大数据市场发展迅速，五年 CAGR 约为 19.7%，增速领跑全球。

用数据治理赋能城市治理、提升城市效能，可以说是近年来商政企的共同愿望。然而，数据治理建设道路依旧充满了挑战，比如数据孤岛、数据风险以及如何有效监管等。

怎么管理？如何共享？

数据孤岛方面，就政府本身而言，构建数字政府打造的政务一体化平台，链接政府内部各部门、各层级，促使政府内部信息交流通畅，打破"数据孤岛""数据高塔"的限制，实现多层级、多部门协同联动的"一体化"办公模式，避免复杂繁琐的政务流程，实现政务资源的统筹协调。

数据风险方面，奇安信科技集团股份有限公司副总裁何新飞在 2021 智博会论坛上分享称：我们希望有一种内生的数据安全的定义，能够在数据治理的同时，就已经在考虑敏感数据、重要数据分布在哪些系统，分布在哪些网络，存储在哪些地方，以及它在流动的过程当中对其进行监测和及时发现。

● 奇安信集团副总裁何新飞在 2021 重庆智博会论坛发表演讲

He Xinfei, Vice President of Qi Anxin Group, delivered a speech at the 2021 Smart China Expo forum.

● 360 政企安全集团在重庆馆展示"工业互联网＋安全生产攻防场景"

360 Enterprise Security Group in Chongqing pavilion displays "industrial Internet + security production attack and defense scenarios".

解码智能时代 从中国国际智能产业博览会瞭望全球智能产业（2018—2022）

而针对如何有效监管这个问题，为保障数据所有权、使用权、隐私权等不受侵害，部分城市已经率先颁布了相关条例，实施立法。比如深圳公布的《深圳经济特区数据条例》，成为国内数据领域首部地方基础性、综合性立法；重庆也颁布了《重庆市数据条例》，该条例于 2022 年 7 月 1 日起施行，在数据安全方面结合重庆实际建立健全了数据处理规则和数据安全体系。

全国城市数据治理，正走向纵深

在生活中享受到便捷高效，是全国数据治理工程给百姓带来的最切实体验。

2022 年的政府工作报告提出，加强数字政府建设，推动政务数据共享，进一步压减各类证明，扩大"跨省通办"范围，基本实现电子证照互通互认，便利企业跨区域经营，加快解决群众关切事项的异地办理问题。推进政务服务事项集成化办理，推出优化不动产登记、车辆检测等便民举措。[1]

国家政务服务平台上线两年多时间，已汇聚了 31 个省（自治区、直辖市）和新疆生产建设兵团以及 46 个国务院部门的政务服务事项。提供涵盖电子证照及教育、助残、司法、民政等多领域服务。

而把眼光聚焦到各大城市，数据治理能力的持续升级为百姓生活带来诸多便利。

2018 年，在"新型智慧城市"全新要求下，深圳提出"六个一"的发展目标，即"一图全面感知""一号走遍深圳""一键可知全局""一体运行联动""一站创新创业""一屏智享生活"。同年，深圳

---

1 薄晨棣、申亚欣，人民网，《政府工作报告：推动政务数据共享 扩大"跨省通办"范围》，2022 年 3 月 5 日。

率先推出政务服务"秒批"改革，实现"网上办、马上办、就近办、一次办"。[1]

2019 年，上海通过推进政务服务"一网通办"，平台事项不断丰富。截至 2021 年底，已接入 3458 个政务服务事项，累计服务超过 136.6 亿人次，累计办件近 2 亿件。2021 年全年，"一网通办"网办率达到 77.03%，全程网办率达 69.30%，"随申办"月活峰值超过 1858 万。[2]

重庆也不甘落后。2021 年 12 月，重庆市政府印发了的《重庆市数据治理"十四五"规划（2021—2025 年）》。其中，明确提出，到 2025 年，城市大数据资源中心全面建成，全市数据图谱与城市信息模型基本建成；数据汇聚率不低于 90%；政务数据共享数量不少于 20000 个，公共数据开放数量不少于 5000 个；数据的准确性、时效性、可用性持续提升，数据共享开放水平走在全国前列。[3]

值得一提的是，在 2021 智博会上，京东科技集团携雄安新区块数据平台、南通市域治理现代化指挥中心等，展出了在渝打造的西南地区首个工业园区智慧环境综合管理系统。该系统在京东云强大后台存储、京东智能城市操作系统高效的时空大数据治理能力支持下，将重庆经开区内 334 家企业环保信息汇集共享、一屏统览，并通过大数据分析研判，人工智能模型监测预警废水、烟雾超标系数，及时发现高污染、高耗能、高耗水"三高"企业，科学分派人员现场处置，用科技指导园区节能减排、规避环保事故发生。[4]

可以看到，全国城市数据治理，正走向纵深，其形成的应用场景正百花齐放，而城市的未来之梦也在一步一步地实现。

---

1 第一财经，《一部法、一张网、一条线，深圳数据治理有这些经验》，2021 年 9 月 28 日。

2 吴頔，解放日报，《"一网通办"将推"免申即享"服务》，2022 年 1 月 5 日。

3 梁浩楠，华龙网，《数据治理"十四五"规划来了！到 2025 年重庆将全面建成新型智慧城市运行管理中心》，2022 年 1 月 5 日。

4 向菊梅，重庆日报，《先睹为快 智博会上有这些"黑科技"》，2021 年 8 月 21 日。

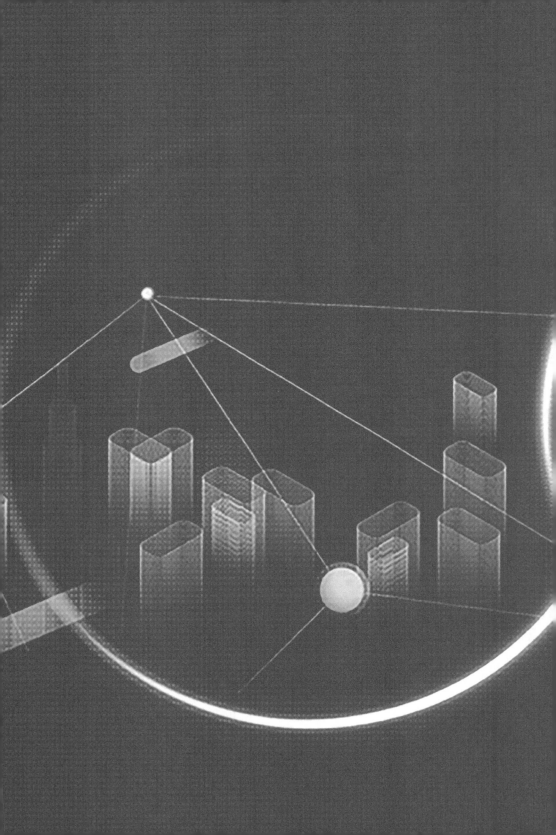

第五章

重庆：一种全新智慧城市的发展探索

　　城市的发展，如同人的成长，时常在激扬的岁月中纵情向前，偶尔也在迷茫的时间点踌躇四顾，一座身处内陆的重工业城市，如何连续创造增长奇迹，又如何在时代的十字路口切换引擎？重庆的蜕变之旅，显然值得思考与总结。

　　而智慧城市的未来又将走向何方？是持续依赖技术的进化？还是取决于自我形成的开放创新生态？重庆也有自己的思考与选择。

第一节
产业升级：从"制造重镇"到"智造重镇"

五年来，从研究峰会到瞭望时代，再回过头来观察这座城市，不知不觉中又产生了新的困惑，也是追问本源的困惑：一个如此清晰的智能产业时代隆隆作响地到来，即便是放眼全球也几乎毫无时间差与信息差，为什么重庆能够迅速形成有竞争力的产业集结？

显然，由于这个问题涉及的变量因素层面之多、领域之广，并非三言两语所能解答。而且，去求索答案的过程，又很容易被提问者逼迫到现实细节的死角，用难以支撑整体的具象化细节，去完成一次象征意义上的解答，从而忽略了这座制造业城市的关键历史。

一个问题还没有答案，另一个问题又会从提问者的脑海中浮现出来：重庆市在智能产业趋势中把握的机会，能否在其他相似城市或异类城市去复制？复制的过程，战略关键点与战术方法论又是什么？

这个问题比上一个更难作答，因为重庆的蜕变仍然是进行时，在一辆持续高速奔行、尚在探索前程的快车之上，在尚未抵达理想中的目的地之前，必须去总结路线和经验，甚至要留下可供后人寻迹的地图，这是难上加难。

对于重庆而言，尝试回答以上问题的过程，既是在躬身一个时代的探索，也是在寻求一个时代的见证。

连续五年的产业升级与增长奇迹

最近五年来，重庆市地区生产总值增加近 1 万亿元、年均增长 6.7%，2021 年达到 2.79 万亿元，人均地区生产总值达到 8.7 万元、高于全国平均水平；供给侧结构性改革持续深化，产业转型升级步伐加快，汽车、电子信息等支柱产业持续壮大；以大数据智能化为引领的创新驱动发展深入推进，"智造重镇""智慧名城"建设取得突破，成功举办四届智博会，数字经济增加值占地区生产总值比重达到 27.2%……[1]

以重庆 GDP 的体量，维持这么高速的增长，实属不易。

重庆是老工业基地，近年来，经济社会发展的各项数据被普遍解读为重庆在经济结构转型方面围绕产业结构调整、推动供给侧结构性改革、加大创新和开放力度取得了比较好的进展。

重庆综合经济研究院院长易小光表示，这几年重庆正大力推进产品结构调整、产业结构调整。过去汽车产业产品结构偏重中低端，现在向中高端转型；电子信息产业中除了笔记本电脑产业之外，液晶面板等其他行业也在兴起。这些行业带动着重庆经济走出低谷。2021年，重庆两大支柱产业汽摩、电子产业分别增长 11.3% 和 17.3%，其中汽车产业增长 12.6%。[2]

中国区域经济研究会副理事长兼秘书长陈耀向认为，东北省份的产业结构调整还处于阵痛期。辽宁的产业中又有一部分是高端装备"大国重器"，不可能像其他产业一样大规模扩张，因此受到一定限制。[3]

1　罗静雯，重庆日报，《牢记嘱托砥砺奋进 五年成就载入重庆史册》，2022 年 5 月 28 日。
2　重庆市统计局、国家统计局重庆调查总队，《2021年重庆市经济运行情况》，2022 年 1 月 19 日。
3　李秀中，第一财经，《四川之后，重庆 GDP 也望超辽宁 西南经济全面超越东北?》，2022 年 1 月 22 日。

重庆的"智变"之旅

那么，重庆经济腾飞的背后有着怎样的密码？

浙江大学国际联合商学院数字经济与金融创新研究中心联席主任盘和林认为，有三点原因：首先，摆脱过剩产业、房地产的 GDP 依赖，着力发展战略新兴产业、实体经济，工业增长成为重庆 GDP 增长的"压舱石"；其次，不盲从东部沿海经验，创新加工贸易模式，让产业转移升级不再是无根之"浮萍"，从而成为重庆 GDP 增长的主力军；最后，社会制度尤其是产权交易创新，如地票制度、城乡统筹推进等，为重庆 GDP 增速带来新动能。[1]

最近几年来，重庆市没有将资源投放到产能过剩产业，做好了房价控制，更集中力量发展汽车、电子核心部件、物联网、高端智能装备、新材料等战略新兴产业、实体经济，使工业增长成为 GDP 增长的主要来源。重庆通过设立基金等引导金融支撑战略新兴产业、实体经济的做法，助力改变这座城市的 GDP 基因。

改革开放以来，特别是中国加入 WTO 之后，扮演"世界工厂"、开展加工贸易，是东部沿海城市经济增长的一股重要力量。但国内传统加工贸易产业多数产业链较短，甚至是原材料与零部件在外，加工完成后销售在外，这种"两头在外"的制造业，利润薄、竞争力弱，其稳定性容易受到原材料、人工成本上升以及海外市场周期性疲软等影响。

而过去几年，重庆创新发展了大规模的加工贸易，从一开始就更注重产业集群的发展模式：一方面延伸产业链，把加工贸易的"微笑曲线"大部分留在重庆，比如生产的每台电脑产值的 70% 都在重庆；另一方面是形成产业聚集，尤其是生产性服务业与制造业的集群，让

1　盘和林，新京报，《重庆凭啥 GDP 增速全国第一？》，2016 年 10 月 28 日。

解码智能时代　从中国国际智能产业博览会瞭望全球智能产业（2018—2022）

● 在极短时间内，重庆形成了世界级的电子产业基地

Chongqing has formed a world-class electronics industry base in a very short period of time.

制造业的底盘更稳。

因此，重庆在极短时间内就形成了世界级的电子产业基地。而随着近年来形成的新能源车、智能汽车产业升级换代，作为国内汽车制造的核心城市之一，重庆市又将这种产业集群的经验用到了汽车制造领域。

2022 年 2 月 28 日，重庆市经济和信息化委员会公布的 2021 年重庆市工业增加值情况显示，规模工业企业产值同比增长 15.8%，其中汽摩产业、电子产业、装备产业增长分别为 11.3%，17.3% 和 15.9%，此外，这三大产业所占全市规模工业的比重分别为 19.8%，28% 和 9.4%。[1]

汽摩产业特别是汽车制造，几乎涉及所有人类已知工业种类，因

1　郑三波，上游新闻，《2021 年重庆规模工业情况来了：企业产值增长 15.8%，全市生产汽车 199.8 万辆》，2022 年 2 月 8 日。

● 三大制造领域在重庆的"补链成群"，已经沉淀下来众多制造业关键技术
The three major manufacturing fields have "supplemented chains in clusters" in Chongqing, and many key manufacturing technologies have been accumulated.

而被誉为现代制造业的明珠；电子制造几乎涉及人类精密制造的极限，因而被誉为现代制造业的皇冠；而全球范围来说，装备制造业由于显著的基石地位而被称为"工业之母"。

汽摩、电子、装备三大产业在重庆市规模工业中的比重高达近六成，充分说明了重庆在先进制造领域布局多年的成就。

这也恰恰是重庆从"制造重镇"向"智造重镇""智变"的底气。

第二节
产业新生：从城市战略到企业重生

2001 年，由于中国在国际社会的两大申请获批，而成为一个极为特殊的年份。

这一年的 7 月 13 日，国际奥委会在莫斯科宣布北京申奥成功，全国各地一片沸腾；而这一年的 12 月 11 日，另一个消息从瑞士日内瓦莱蒙湖畔传来，虽然也振奋了国人，但并没有获得北京申奥成功那样广泛的全民关注。

站在 21 年后的今天，我们当然可以总结，后一个事件无论对于中国的影响，还是对于世界的改变，都可谓"百年难得一遇"，因为这一天，WTO 在总部召开会议，批准了中国的加入。

这个决定，改变了此后二十年来的全球经济格局。从这一天开始，全球经济一体化的速度被加快，中国迅速成为世界工厂，中国经济也正式融入全球经济圈，并在全球一体化的进程中发挥越来越重要的新增长引擎作用。

从性价比到源创新

加入 WTO 的第一个十年，中国制造在源创新方面未必领先，但将控制生产制造成本的能力发挥到极致。这是中国在全球范围内变身世界工厂的最核心动力源。而加入 WTO 的第二个十年，中国制造已经不仅限于扮演世界工厂的角色。得益于中国庞大的消费市场与活跃的创新意识，大量的中国制造型企业，逐渐参与全球创新并在众多领域开始引领商业变革。

十几年来，中国智能手机的成长路径几乎就是中国式创新的一个完美诠释模板。

在各大国际巨头林立的智能手机市场中，过去十年，中国手机厂商集体杀出了一条血路。小米手机在移动互联网爆发式增长的大趋势中，快速整合全球手机产业链的最新技术，并以极致性价比获得消费者的信任，从而获得了巨大成功。这种竞争策略，迅速影响了中国手机厂商，并成为整个智能手机行业的集体兵法，因此形成了在中国市场特有的手机产业格局。除了 iPhone 能够凭借独树一帜的创新能力在中国市场长期立足，排名靠前的手机品牌已经清一色地被中国品牌所取代。不仅如此，甚至它们顺势走出国门，成功占领了相当比例的国际市场。

重庆制造在国内市场的地位与中国制造在全球市场的地位，在基本逻辑上恰好相似：制造业基础环境相对成熟，人力成本等相较沿海更低，转换新技术成为新产品的速度也比较快。因此，在中国成熟的制造业体系内，重庆很容易复制中国制造在全球市场的成功。

从摩托车制造时代开始，重庆制造就很善于借鉴日韩摩托车产业的经验，卡位低成本、追求高效率，将已经被验证过的新技术制造成新产品，并将之以较高的性价比提供给消费者。而到了汽车消费时代，这种竞争优势更是发挥到了极致。

2021 年夏天，本书创作团队走进长安汽车，令人印象深刻的是，一位负责技术方面的管理人员在接受采访时表示：长安的核心竞争力其实在于将已经被市场验证过的先进技术，速度更快、成本更低地带给消费者。

● 长安汽车 5G 智慧工厂应用场景
Chang'an Automobile 5G Smart Factory Application Scenario

某种意义上来讲，这也是重庆制造在很长一段时间内差异化生存的根本。无论是在笔记本电脑、手机等制造业西迁的过程中，还是在众多汽车生产制造厂商在重庆陆续聚集的过程中，都体现了这一个根本的竞争力。

然而，就像中国制造在全球市场要升级为中国创造，重庆的制造业也必然要经历从性价比到源创新的转变。

长安汽车的触底反弹

2018 年，中国汽车市场在宏观经济下行、居民消费能力减弱、汽车保有量较高以及此前的购置税优惠政策透支等多种原因的综合影响下，在经历了 28 年高速成长之后，销量首次出现负增长，同比下滑 3%，次年下滑幅度再度扩大到 9.6%。

自此，中国汽车市场进入一个"全新时代"。

正是在这种情况下，总部位于重庆的长安汽车在这段时间内经历了断崖式的"三连跳"：2017 年，长安汽车累计销量 287.2 万辆，同比下滑 6.23%；2018 年其累计销量为 213.78 万辆，同比大幅下滑 25.6%；2019 年全年销售量 175.9 万余辆，同比再次下滑 15.16%。2019 年，长安汽车亏损高达 26.47 亿元人民币，是其上市 23 年来首次出现年度亏损。[1]

长安汽车的失速三连跳，不但震惊了整个中国汽车制造领域，更是引发了重庆汽车圈的一场深思。

要知道，在长安汽车的"三连跳"之前，重庆汽车在全国可是"三连冠"。2016 年，重庆汽车产量高达 316 万辆，位居全国各省市第一，在全国占比超 11%，连续三年蝉联全国省市汽车产量第一，也是全国唯一一个汽车年产量超过 300 万辆的省市，堪称名副其实的"汽车之都"。[2]

而作为重庆汽车的核心龙头厂商，长安汽车的断崖式下滑，直接终结了重庆的桂冠之旅。

无论是一家汽车厂，还是一座汽车城，遭遇如此严重的断崖式下滑，要想迅速地恢复起来，想一想都是一项难以达成的目标。因为下滑背后，意味着一家龙头厂商、一个产业体系、一座汽车城市竞争力整体退化的长期积累，甚至是三种问题叠加的结果。

在这种情况下，若想触底反弹，除了要找到企业、产业的病因，寻找新的增长契机，还要考虑市场变化、惯性的品牌落差、自然的市场响应周期等复杂因素，这是极为不易的。

整个汽车圈都认为，长安汽车完了，重庆地位不保。

然而，长安汽车让整个行业再次震惊了。2019 年以后，长安汽车在新一轮的产品攻势下，迎来了增长周期，销量触底反弹。

---

1 澎湃新闻，《长安汽车复苏时刻，朱华荣接棒张宝林出任董事长》，2020 年 6 月 11 日。
2 蒋艳、周雅丽，重庆晨报，《年产整车 30 万辆 北京现代重庆工厂落成》，2017 年 7 月 19 日。

● 长安汽车之所以能触底反弹，在于坚持创新研发新车型

Chang'an Automobile has rebounded from the bottom with its persistence on innovative research and development of new models.

2020 年，长安汽车全年实现整车销售 200.4 万辆，同比增长 14.0%，重新突破 200 万辆大关；2021 年，长安汽车继续增长，旗下各品牌全年销量高达 230 万辆，同比增长 14.8%。与此同时，长安汽车的营业收入与净利润同样节节攀升，扭亏为盈，持续盈利。

时至今日，长安汽车的这种势头仍未消失，2022 年 3 月，长安汽车的批发销量为 132483 辆，同比增长 20%，超越一汽大众、上汽大众，再次冲到了国内厂商的第一名。[1]

长安汽车触底反弹的强劲动能背后，经历了颇为关键的转型。

1　数据来源：中国汽车流通协会汽车市场研究分会《2022 年 3 月份全国乘用车市场深度分析报告》。

持续内化城市的智能化红利

2017 年，察觉到自身不足（销量和业绩过度依赖合资品牌、市场增长缓慢）的长安汽车开启了第三次创业计划，宣布向智能出行科技公司转型，开始聚焦"新能源化""品牌向上"和"智能化"三大层面。

当时，在外界看来这是长安汽车的一次冒险之举，毕竟"船大难掉头"，何况未来之路暗礁遍布，风云难测，也给长安汽车增加更多潜在风险。为了完成转型，从 2017 年开始，长安先后启动了新能源战略"香格里拉计划"，智能化战略"北斗天枢计划"，客户的"一号工程"数字转型工程等，体系能力开始逐步提升。与此同时，基于长安蓝鲸 NE 动力平台打造而来的蓝鲸动力总成陆续得到应用，让长安汽车在近几年的市场竞争中脱颖而出。

此外，高端新能源品牌是当前车企争相布局的热门领域，无论是蔚来、理想、小鹏、恒大等造车新势力的闯入，还是从去年开始岚图、极狐（ARCFOX）、智己等国内车企集团的加码，都让该领域的热度不断飙升。当然，仍处于重要转型期的长安汽车也不甘人后。

2020 年 11 月，长安汽车官方曾宣布正在携手华为和宁德时代三方联合打造高端智能汽车品牌，其中包括一个智能电动汽车平台、一系列智能汽车产品和一个超级"人车家"智慧生活和智慧能源生态。

企业创新离不开持续的研发投入，研发投入是企业实现创新的物质基础和保障。近年来，长安汽车持续将年销售收入的 5% 投入研发当中。据长安汽车财报透露，2019 年全年，长安汽车投入 44.78 亿元用于产品和技术项目开发[1]，研发投入占营业收入比例为 6.34%，在行业中算是较高水平。

---

1 刘勇、闫涵，上游新闻，《长安汽车一季度盈利 6 亿元 今年力争产销汽车超 191 万辆》，2020 年 4 月 30 日。

未来，围绕着智能化领域，长安汽车将持续加强对软件能力的建设。按照规划，2025年长安汽车将打造5000人规模智能化开发团队，软件开发人员占比将提升至60%。

从2020年开始，长安汽车迎来了两大反转，一个是表面上广为人知的销量反转，一个是骨子里深刻内省的理念反转。

从重庆汽车制造的这段短暂的跌宕，可以总结出两个事实：

一个是重庆制造的超强自我修复能力，即便是充满了各种不可能，仍然实现了下跌后的反弹。另一个是在创新能力缺位下的性价比蕴含着巨大的风险，当消费市场的需求产生风吹草动，当技术创新的巨浪改变了比赛规则，当一座先进制造业城市的发展节奏与全国甚至全球出现脱节，其后果是多么严重。

在2021智博会上，长安汽车董事长朱华荣在发表演讲时有感而

图片来源：郑宇/视觉重庆
Photo by: Zheng Yu/Visual Chongqing

● 长安汽车董事长朱华荣在2021智博会上发表演讲
Chairman of Chang'an, Zhu Huarong delivered a speech at the 2021 Smart China Expo.

发："近三年来我们投入近百亿，目前掌握了 500 余项智能低碳技术，其中 48 项是中国首发。我们投入了智能网联汽车 145.6 万辆，在全球处于领先地位。"

发展自身的核心竞争力、引领先进的科技实力，升级全新时代的重庆制造，甚至从长远来看，实现一座城市整体意义上的智能化、智慧化，显然已经成为这座城市深度思考、长远构思的核心任务。

长安汽车与众多的重庆智造型企业一起，正在将一座城市的智能战略红利，持续内化为企业自身的竞争力。

第三节
谋定后动：重庆智慧城市的发展路径

一座城市，拥有自己的生命，也拥有自己的坚持与时代的运势。

我们回首那些屹立全球的著名城市，似乎都能追溯到一个类似马孔多的小城起点，而在这些城市的历史中，往往能找到一个基于特殊历史机遇而快速进化的高光时刻。

重庆，在不同的历史时代，数度转型中，都没有迷失自身的制造业能力，也终于等来了工业制造与智能产业相交融的时代。

一座梦想之城的起点

在 2017 年之前，重庆 GDP 高歌猛进的同时，也有产业布局蹉跎的一面。

它在很长一段时光里，已经多次错过了战略蜕变的机遇。

由于身处西部内陆，并没有在中国加入 WTO 之后获得过多的沿海运气；由于缺乏产业基础，并没有在移动互联网的浪潮中抓住多少风口；由于本地制造以重工业为主，由小商品供应链为主推动的电子商务蓬勃中也几乎和它没什么关系。

而从 2017 年开始，重庆又是幸运的，终于迎来了一个可以从更前瞻的视角、更长久的耐心、更果决的实操，为重庆这座城市谋划一个面向未来的机会：作为一座以制造业著称的城市，重庆未来的机会只能是以人工智能为制造业做加法，应该更大胆地拥抱智能时代，以人工智能去武装一个拥有丰富制造业基础资源的城市，使之完成从传统制造向智能制造的关键升级。

从"制造"到"智造"，表面上看起来是一字之差，却有天壤之别。

因为这样的机会，重庆这座城，感受到了一个全新时代的到来、一股全新力量的积聚、一种崭新生命的重组。

智能时代来了，谁能想到重庆会有这么大的勇气，将自己置身于一个最前沿领域的风口。

无数的人工智能标杆企业、创新公司，在创立之初都是盯着北上广深等城市，因为最前沿的人工智能技术，需要与最新鲜的咨询、最优秀的人才、最活跃的市场相匹配。而当他们仔细打量过重庆这座地处西部的制造业城市之后，才猛然发现有关于智能制造的一切缥缈而富有创意的想象，在这里都能得到最直接与真实的制造业资源。

通过 100 年来的工业积累，重庆已经成为全国范围内制造业品类最丰富、供应链上下游配套最齐全的工业城市。再加上近年来重庆的 IT、电子、软件、装备等制造产业也走在国内前沿，而且还是全国工业互联网标识解析五大顶级节点之一。由此形成的工业互联网成熟条件，足以使任何企业与团队侧目。

于是，从 2018 年开始，围绕智博会的舞台和展台，重复上演了无数的故事：

一个又一个智能领域精英，从智博会的舞台上走下来，迫不及待地在摩天大楼中开始寻找办公室；一家又一家人工智能企业，从博览会的展台中走出来，迫不及待地去工厂车间里寻找合作伙伴。

在他们的背后，是一批又一批与重庆偶然相遇、必然落地的企业，也是一次又一次与世界同频共振、顺势崛起的机遇。

解码智能时代 从中国国际智能产业博览会瞭望全球智能产业（2018—2022）

● 从 2018 年开始，重庆搭建起了智博会的大舞台
Since 2018, Chongqing has set up a big stage for the Smart China Expo.

　　或许在重庆本地人后知后觉的情况下，重庆已经成为全球智能产业的目的地。

　　有时，城市的关键选择，会给历史留下重重的印记。与这些时刻相比，曾经的颠沛与流离，则显得没有那么重要了。

　　重庆正在把自己的制造业生命力从工厂车间里释放出来，在展台上，在演讲台上，在数字世界等各种空间挥洒，智博会就是重庆这种生命方式的一个重要突变机会。

　　在当代，一家企业、一种产业、一座城市，其长远生命力的衡量标准不再是简单的利润、营收和 GDP，而是要加上技术的变量去激活智能制造的产业底盘，进而持续输出面向未来的竞争力。

智能产业时代的重庆想象力

在山水相依、立体交错的晨雾中，在鲜香扑鼻、麻辣交融的夜色里，来自天南海北的人们，正在以自己的认知为重庆这座城打上一个又一个全新的标签。

而在今时今日重庆的各种标签中，有一半竟与智能有关。

因为重庆在经历了近一百年的制造业积累之后，终于等来了以大数据、智能化为基底的人工智能浇灌，一个是硬核的骨骼，一个是软性的血液，二者的结合立即形成了数字时代的都市生机，而重庆的发展也愈发清晰：目标是打造一座有血、有肉、有机、有生态的"智造重镇"，也是一座更美、更绿、更宜居的"智慧名城"。

长期以来，关于到底什么是智慧城市，其实每个城市都有自己不同的理解、诠释、愿景与实践，而对于立志打造"智造重镇"与"智慧名城"的重庆而言，一个清晰而完整的智慧城市，已经呼之欲出。

简而言之，"智造重镇"与"智慧名城"就是重庆智慧城市战略的一体两面，分别从城市经济与城市生活两个角度，去构建重庆智慧城市的未来格局。

按照这个思路来复盘重庆的智慧城市推进步骤，不难发现科学务实之处。简单总结，可以分为三步走：分别从经济发展、治理结构、社会生活三个角度，构建"芯屏器核网"全产业链、"云联数算用"全要素群、"住业游乐购"全场景集，逐步推进经济结构数字化、基础设施数字化、生活应用数字化，战略层次清晰，推进节奏合理，更符合智慧城市的本质追求。

而在这个整体战略的推进过程中，智博会扮演了尤为重要的角色，它既是打开天眼瞭望世界的窗口，也是敞开心胸接纳未来的门户。智博会的持续主办，将伴随重庆智慧城市战略不同阶段的逐步推进，源源不断地提供关键支撑。

2022 年是重庆举办智博会的第五年，也是重庆面向一个前沿产业深度对接的第五年。智博会的大门一开，声音传出去，产业领进门，随着一家又一家在细分领域拥有无限可能的创新企业，将自己的人工智能总部落地重庆市，这座城市的智能制造产业生态已经初具雏形。

　　至此，一硬一软，一刚一柔，智能产业的骨血相融已经齐备。那么，重庆市的智能制造基因，也就完成了顺应时代的自我突变。前面就是开阔的旷野，虽然也会有自我蜕变的艰难，但不会再有何去何从的迷茫。

第四节
智慧城市：重庆的未来还会怎样？

过去，人们更习惯于以经济水平、治理水平、人文环境和平均收入等指标，来衡量一座城市的发展水平。而这些指标本身将继续存在，用以衡量城市在区域、全国及全球的地位。

但是，进入智能时代，探索智慧城市将成为绝大多数城市的一个全新共同目标。如何以新一代信息技术实现信息化、工业化与城镇化深度融合，缓解"大城市病"，提高城镇化质量，实现精细化和动态管理，并提升城市管理成效和改善市民生活质量，成为全球城市"智慧城市"建设的共同使命。

关于重庆，我们讲了很多现在，那么未来又会怎样呢？

智慧城市的未来，不仅是智能城市

可能很多人会好奇，这个地球上到底有多少个城市？

显然，受制于不同国家的经济发展、聚居习惯、行政区划、城市标准等天然差异，每个国家对于城市的定义都有所差异，这个问题的答案甚至很难用常规统计方式去寻求。

而怀有这种好奇心的同济大学吴志强院士团队，从 2014 年便开始用人工智能的方式解答这个问题，到了 2018 年 1 月终于有了答案：

解码智能时代

从中国国际智能产业博览会瞭望全球智能产业（2018—2022）

全球面积在 1 平方公里以上所有的城市和建成区共 13810 个。[1]

对于全球所有城市，吴志强院士团队不仅仅是进行了统计，而且把这 13810 个城市树的曲线边缘精确地画了出来。2014 年起，吴志强和团队开始研发"城市树"全球影像智能识别技术，就是通过 30m×30m 精度的网格，将 40 年时间跨度内的全世界所有城市的卫星遥感图片进行智能动态识别并叠加，因得到的城市时空演进可视化轨迹呈树状，所以命名为"城市树"。

面对着全球首次被统计出来的这 13810 个城市，吴志强院士团队利用人工智能技术，甚至建立了世界城市演进数据库，运用人工智能图像识别技术，寻找城市建成区范围，总结世界城市发展规律，甚至可以推演城市 2020 年、2035 年、2050 年的发展情景，诊断城市的未来问题，为研究整个世界的人类城市问题做支撑。

随着人类社会进入智能时代，有一个时代的思考题，摆在了全球 13810 位市长的面前：如何将自己的城市建设成为智慧城市？

在 13810 位市长们的脑海中对于智慧城市的想象或许不是相同的答案，但不难在近十年来全球智慧城市的探索过程中总结出目前已经形成的部分共识。

智慧城市绝对不仅仅是智能城市。

智慧城市能够通过各种信息技术和智能化应用，为经济的发展、城市的治理、市民的生活，提高效率、提升便捷，但又不能止步于此。智慧城市需要人工智能技术的支撑，但不能把整座城市的命运交给机器。如果只是单纯将各种物联网设备进行自动化，按照人工智能程序设定运行，以及将各种仪器嵌入城市的各个角落，那么这座城市也只能沦为一个巨大的毫无生气的机器，徒具智慧城市的外形，而非智慧城市的本质。

---

1 张炯强，新民晚报，《同济大学吴志强团队绘制全球首张最完整的 13810 个城市建成区全图》，2019 年 8 月 12 日。

● 西部（重庆）科学城首届场景大会，凸显"人"是整个智慧城市的核心
The first conference of Western China (Chongqing) Science City highlights that people are the core of the entire smart city.

人类智慧的加入，对于智慧城市而言更为重要。

之所以称为"智慧城市"，而不是"智能城市"，显而易见的区别在于，人才是整个智慧城市的核心。城市服务于居住于其中的人，人根植于充分科技化的城市。人工智能技术可以简化掉那些烦琐、机械、低效、高耗的城市运行方式，剩下的空间则应留给人们个性生活、独立思考。

智慧城市，重心在开放创新生态

一直以来，很多城市推动智慧城市战略，往往重心会落在以物联网、云计算、移动互联网为代表的新一代信息技术，盲目追求各种智慧城市的项目建设，而忽略了城市的智能产业发展，这种智慧城市即便建设起来，也很难形成自己的"城市生命"，拥有面向未来的进化

解码智能时代
从中国国际智能产业博览会瞭望全球智能产业（2018—2022）

能力。

在智慧城市的智能技术硬件之下，知识社会环境下逐步孕育的软件实则更为重要，即不断为智慧城市的迭代而贡献智慧的创新型企业与创新型人才，以及整个城市为其打造的孕育环境。

智慧城市应有自己的开放创新生态。

因此，智慧城市的最终形态，并不是一群建筑和一堆硬件，不仅依赖各种技术的进步，也需要社会的发展来进行推动，形成一种不断演进、不断成长的形态。

同时，创新型企业的引入与扎根，需要创新型经济的肥沃土壤；创新型人才的吸纳与成长，也需要创新型产业的充沛光照。重庆市在智能时代来临之际，果断地迈向智能制造，为经济发展打下关键基础，更将源源不断地为智慧城市建设提供创新动力。

以集成电路设计业为例，2020 年，重庆市集成电路设计业销售增速达 206.1%，有 3 家企业实现销售过亿元；全市"芯""屏"双核同比增速均超过 20%。[1]

而在集成电路设计业高速发展的背后，是不同产业之间的相互联动、相互赋能，重庆市拥有 2 亿台智能终端、6000 万台笔电、3000 万台家电、200 万辆汽车的市场需求；"芯屏器核网"全产业链、"云联数算用"全要素群，"住业游乐购"全场景集，更需多样化的半导体技术、标准、产品支撑。半导体产业已成为重庆产业链条中应用领域最广、渗透范围最深的产业，被誉为中国 AI 四小龙之一的云从科技，也孕育于此。

过去五年，对于重庆而言，取得的成就，终将载入史册。

重庆市地区生产总值五年增加了近 1 万亿元、年均增长 6.7%，

---

1 郑三波，重庆商报，《集成电路设计业 去年我市增速 206.1%》，2021 年 1 月 18 日。

2018 年进入两万亿俱乐部，2021 年达到 2.79 万亿元，人均地区生产总值达到 8.7 万元，高于全国平均水平，而 2022 年也有望再创新高；供给侧结构性改革持续深化，产业转型升级步伐加快，汽车、电子信息等支柱产业持续壮大。以大数据智能化为引领的创新驱动发展深入推进，"智造重镇""智慧名城"建设取得突破，成功举办四届智博会，数字经济增加值占地区生产总值比重达到 27.2%；从国家层面把握机遇，成渝联动规划双城经济圈，一批重要规划方案的编制出台、成渝中线高铁的标志性重大项目的顺利实施、万达开川渝统筹发展示范区等区域合作平台加快建设、成渝金融法院获批设立等众多成果，已经为打造未来中国区域经济的第四极埋下了关键伏笔。

追溯这些成就的源头，不难发现五年前的那个关键节点：一座地处中国西部的重工业之城，在一个全新时代到来的时候，坚定地推开了智能时代的大门，勇敢前行。

五年过后，整个世界发现，重庆这座城市已经深深地打下了智能时代的烙印，无论是一年一度的智博会，还是高度自动化的智能工厂，以及散落在这座城市人们眉宇间的自信、火锅旁的谈资、会议室的探讨，都是智能时代的清晰印记。

而未来五年或是更久远的将来，这座已经与智能时代深度绑定的城市又会怎样？这已经是所有关心这座城市未来的人，都密切关注的一个问题。

8 月 24 至 25 日，2021 智博会举办期间，礼嘉智慧公园推出了"智慧生活的一天"，提供了一个让参观者提前感受未来生活的空间。

"智慧生活的一天"应用场景，是礼嘉智慧公园以日常一天活动空间转换逻辑，按照身临其境、真实体感标准打造的集智慧起居、智慧生活、智慧工作、智慧医疗、智慧艺趣等主题为一体的大型活动项目，包括 60 个体验场景、130 个体验项目，充分展现智能化在高品质生活中发挥的引领作用。

图片来源：张锦辉／视觉重庆
Photo by: Zhang Jinhui/Visual Chongqing

● 2021 智博会举办期间，礼嘉智慧公园推出了"智慧生活的一天"
During the 2021 Smart China Expo, Lijia Intelligent Park launched "A Day of Smart Life".

在这 130 个体验项目中，"智慧生活的一天"充分以人为中心，用最先进的物联网与大数据技术，构建起了既充满科技智能，也兼具人性洞察的智慧生活。

那么，重庆的未来就是这样的吗？或许还不是。信息技术、数据算法与硬件创新，在过去数十年间，总是以颠覆人类想象的方式，不断创造出全新的智能产业物种，我们今天的想象，仅限于已经理解的方式，并不能完全代表未来。在人类关于未来的想象力边缘之外，拥有更丰富的人工智能创新空间，而在真正的智慧城市呈现在人们面前那一天，将收获无尽的惊喜。

但对于重庆而言，这已经是一个很好的开端，身处智能产业时代之中，逐步构建属于这座城市的数字化经济、数字化治理和数字化生活，必将沉淀下越来越多的经验与资源。

而重庆的未来，则构建于所有这一切探索之上，值得想象，值得自豪，值得相信。

后记

　　为了呼应智能时代发展的新趋势，展示智能产业取得的新成果，同时也为广大群众打造一套了解智能时代、融入智能时代的优秀科普读物，2020 年，中共重庆市委宣传部决定持续性地出版"解码智能时代丛书"，并明确要求以国际化的标准，将"解码智能时代丛书"打造为智博会的一张文化名片。

　　此次出版的 2 种图书即"解码智能时代"这一伴智博会而生的"文化名片"的生长和延续。本丛书整体工作由中共重庆市委宣传部统筹组织。中共重庆市委常委、宣传部长姜辉同志指示，要围绕重庆市"智造重镇""智慧名城"建设总体战略，围绕这五年智博会及智能产业的成就，做好今年"解码智能时代丛书"的出版工作。

　　本丛书共图书 2 种，其中《解码智能时代：从中国国际智能产业博览会瞭望全球智能产业（2018—2022）》（中英双语）以图文并茂的形式，梳理全球智能产业发展的脉络，展现历届智博会的精华，回顾重庆市智慧城市建设的成果，并对重庆在智能产业时代的未来发展、智慧城市建设的未来实践进行分析与展望；《解码智能时代：重新定义智慧城市》展现了大数据智能化如何推动城市有机更新，积极拓展特色智慧应用场景，探索国际化、绿色化、智能化、人文化美好城市的实践路径。

在整个编写及出版过程中，中共重庆市委宣传部常务副部长曹清尧同志对本丛书进行了全面指导，出版处统筹多个专业团队紧密协同，并对丛书的策划创意和内容质量进行总体审核把关，有序推动丛书的编写及出版工作。本丛书的调研、写作及出版，还得到了重庆市经信委、重庆市大数据应用发展管理局以及智博会秘书处等部门的大力支持。

重庆大学出版社特别邀请智博会秘书处何永红主任、重庆大学大数据与软件学院向宏教授、重庆大学管理科学与房地产学院曾德珩教授对系列读物进行了审读，重庆大学出版社饶帮华社长组织了多名资深编辑对书稿进行字斟句酌的打磨，确保内容的科学性、可读性、准确性。

如果站在智能时代历史进程的维度上，我们希望"解码智能时代丛书"能够以年度为单位，记录与展望智能化究竟如何为经济赋能、为生活添彩，记录与展望"数字产业化、产业数字化"的实践过程，记录与展望人类文明史上这场伟大而深刻的变革。这样的记录与展望，这样的智能时代年度印记，是有历史意义的。

谨此，致敬智博会，并对所有促成本丛书立项、提供写作素材、从事书稿编写与翻译、参与本丛书审订、帮助本丛书出版的单位与个人，对接受写作团队采访的专家，致以深深的谢意。

编写组

2022 年 7 月

解码智能时代

从中国国际智能产业博览会瞭望全球智能产业（2018—2022）

DECRYPTING
THE INTELLIGENT ERA

Overlook the Global
Intelligent Industry from Smart China Expo

（2018—2022）

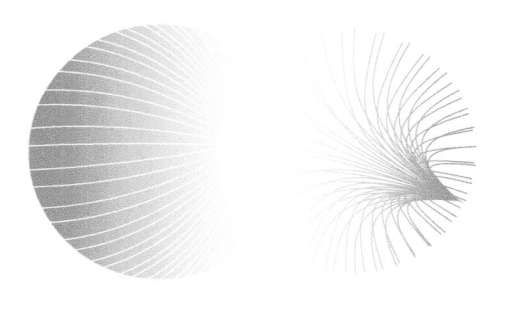

重庆大学出版社

"The world has entered a period of rapid growth in the digital economy, in which new technologies, business patterns and platforms such as 5G, artificial intelligence and smart cities have sprung up and greatly influenced scientific innovation, industrial structure adjustment and economic and social development across the globe.

In recent years, China has actively promoted digital industrialization and industrial digitization to push for deep integration between digital technologies and economic and social development

On the occasion of the 20th founding anniversary of the SCO, China is willing to, together with other SCO member states, carry forward the Shanghai Spirit and get deeply involved in international cooperation on the digital economy. Digitization, networking and intelligence shall provide more momentum for economic and social development, breaking new ground in digital economy cooperation."

— Excerpted from Chinese President Xi Jinping's Congratulatory letter to the China-Shanghai Cooperation Organization (SCO) Forum on the Digital Economy Industry and Smart China Expo 2021
(Xinhua News Agency, Beijing, Aug. 23, 2021)

CONTENTS

目录

Part III
Intelligent Life: Intelligent Core for Life

Part IV
Intelligent Governance: Building an Intelligent Hub for Smart Cities

Part V
Chongqing: An Exploration of a Brand-New Smart City

Afterwords

Part I

Five-Year Smart China Expo: Witnessing the World from One Era to Another

All of a sudden, the global science and technology community is somewhat lively—a kind of unreal but colorful era scenario, a kind of unclear but deafening sound of the times, swept here, filling the world. Like flags covering the sky, like drum and horns chirping in unison, like horses galloping, like all the musical instruments playing together, it makes people feel an inexplicable exhilaration all over their body, and then actively involve themselves in it without hesitation. In the past five years, in this vigorous process, artificial intelligence(AI) has rapidly become the common topic of all countries, cities, enterprises, and experts facing the future world. The five-year Smart China Expo brings together the power and imagination of the entire smart industry, and thus witnesses the initiation and formation of an intelligent era.

Section I

The Evolution Epitome of the World in the Five-Year Artificial Intelligence's Development

When we stand in 2022 and look back at 2018, everyone will have different memories. In just five years, a careful comparison of the changes therein inevitably gives rise to a sense of vicissitudes.

Do you still remember how the sharing economy was in full swing? Shared bicycles, shared cars, shared B&B, shared chargers and flexible labor services suddenly became popular. China's sharing economy has quickly become the innovator and leader of the global sharing economy because of its wide coverage of transportation, housing and accommodation, knowledge skills, life services and many other fields.

Do you still remember the emergence of new retail? Relying on a decade of booming e-commerce development in China and the reverse penetration of the Internet into traditional offline commerce, the tentacles of Internet retailing began to extend offline in various innovative forms, with cashierless stores, food delivery, supermarkets to homes, community group purchases, home delivery of medicine, Internet grocery shopping and many other scenarios. A vigorous new retail revolution in China has become the innovation source of the global digital economy because of its rich imagination, courageous innovation and huge market.

These dazzling explorations of China's new economy have gone from jerky embryos to mature models. Since 2018, China's innovation has finally started to lead the world in some areas.

Around 2010, Chinese innovators seized the innovation opportunities of mobile Internet, synchronized with the global technology trends, and stood at almost the same starting point as global innovators. Relying on the booming domestic market, we could enjoy a unique dividend

of China's consumption upgrades. During this period, Chinese entrepreneurs no longer copied foreign mature models as they did in the early days of the Internet, but more and more made original innovations in China, and began to export them to the world, spreading their influence to the whole world.

The focus of this book is from 2018. Under the background that China's innovation was popular all over the world, the global economy was brewing to switch the main engine of innovation. The five-year journey of AI has moved from laboratory to the new era of intelligence at all levels of global economy, life and social governance.

The times created an ingenious world, and the city started to follow the trend. Let us start from the beginning of 2018.

China's Innovation Began to Affect the World

In May 2017, a Romanian graduate student in China named Peter greatly inspired the pride of many Chinese people. In a media interview, he talked in front of the camera about what it was like to live in China with excitement, "I am obsessed with the Four Great Inventions, and now China is developing so fast that there are already 'New Four Great Inventions'".

The "New Four Great Inventions" came from a survey initiated by the Silk Road Research Institute of Beijing Foreign Studies University. Students from 20 countries along the Belt and Road voted for the "most important Chinese lifestyle that they would like to bring back to their countries". At the top of the list are high-speed railway, online shopping, mobile payments and shared bicycles.

Strictly speaking, none of the "New Four Great Inventions" is invented in China from a technical point of view. However, the application of these four innovations in the Chinese market is far ahead in the world in terms of scenario construction, model innovation, and the scale of application, therefore, it is not an overstatement that they are named as China's "New Four Great Inventions".

Along with China's rapid economic development and coinciding with the rapid global transformation of mobile communication technol-

ogy, China has become a hot destination of mobile Internet innovation, and Chinese consumers are among the most active in embracing, experiencing, and participating in innovation worldwide.

Let's start with mobile payments. The Japanese invented QR codes first, but the contribution of China to the global market in using QR codes for mobile payments and pushing it from a technology to a large-scale application for the whole society is huge.

Since 2013, after the fierce competition in the market of online ride-hailing and bike sharing, the popularity of mobile payments in China has been amazingly fast, and in just a few years, even the aunts selling fruits at roadside stalls are using QR codes to collect money. In 2016, the size of China's mobile payment market reached 157.55 trillion yuan. According to estimates from global consulting firm Forrester Research, in 2016, the size of the mobile payment market in the United States was $112 billion, a 200-fold difference in size between the two.[1]

Speaking of food delivery, Doordash, the earliest food delivery platform in the United States, was established in 2013, which immediately inspired Ele.me, which had been established for 4 years, and Meituan, which had been established for 3 years. They quickly tapped the food delivery market and launched a full-scale competition in the Chinese market with almost no time difference.

By 2017, China's food delivery market had become a prairie fire. After many small platforms were eliminated, Meituan and Ele.me began to compete fiercely, each forming a massive team of food delivery riders. When Chinese consumers ordered food at midnight, they could get it in half an hour. While almost all foreign take-away platforms did not have their own delivery capabilities that year. It took them a few years to finally learn how Chinese companies built their own distribution networks after 2020.

Another example is bike sharing. Social Bicycles, Inc. in New York was established in 2010 (renamed JUMP in 2018 and acquired by Uber in the same year), but initially did not actually participate in the operation of public bicycles, but as a technology supplier, selling technology

---

1   Li Yanxia, *Xinhuanet*, "Central Bank Report: China's Mobile Payment Amount Increased by Nearly 50% Year-on-Year in 2016", March 16, 2017.

to the operator. In 2013, the company launched an independently-operated shared bicycle. Although it has developed a GPS-enabled locking system integrating GPS positioning and PIN code unlocking from the technical level, in order to prevent users from parking indiscriminately, the company still set up physical stops, which led to the increase in operation cost, so the early development speed was relatively slow. Ofo, founded in Beijing in 2014, aimed at the GPS-enabled locking system sharing solution from the very beginning. It started Angel financing from the following year, and the product was launched in one go. In 2016, it achieved the financing from the pre-A round to the C round within half a year. Since then, the development speed of the two markets in China and the United States in the field of bike sharing has shown a world of difference.

Starting in 2018, short videos and livestreaming e-comerce with distinctive Chinese original features have gone viral, becoming a role model for global innovators to emulate. In that year, TikTok, the international version of Douyin, was launched and ranked first in major application markets for several years. It was not until 2021 that major American companies such as Amazon and Google began to invest heavily in livestreaming e-commerce. In the past 2020, the market size of China's livestreaming e-commerce had doubled from the previous year to nearly 1 trillion yuan.

When we look back at these mobile Internet innovations of the past, it is easy to see that 2017 was a unique time point. Prior to this, China's digital economy developed rapidly, but under closer inspection, it often lacked originality. Starting from 2017, China's innovative companies, began to reveal more and more originality, and China's digital economy undertones, gradually evolved from imitative innovation to original innovation.

We can never underestimate this change because it will determine the core creativity of a country and a market, based on trends and facing the future, especially when the whole world is about to enter a brand-new era of intelligent industry. How much energy can Chinese innovation create in the future?

The Wonderful Debut of Artificial Intelligence

On May 23, 2017, Ke Jie, a ninth-degree Chinese Go player and the world's top-ranked Go champion, quickly became an Internet sensation when he played against AlphaGo, an AI program developed by Google's company DeepMind, in Wuzhen, Zhejiang province.

The AI chess player challenged the talented human chess player, and the ending was already doomed, but those symbolic moments were still unforgettable.

Just a year ago, the old version of AlphaGo, powered by 176 Nvidia chips, had a man-machine Go race with Lee Sedol, the world Go champion and professional ninth-degree player, and won 4:1 with the application of new technologies such as neural network and deep learning. After defeating Lee Sedol, AlphaGo quickly got a new upgrade, and even before challenging Ke Jie, it had already registered as "Master" on the Go website, playing against dozens of top human Go masters in turn, achieving a brilliant record of 60 wins and 0 loss.

The AlphaGo that Ke Jie faced is the upgraded version of Master.

AlphaGo defeated Ke Jie by a total score of 3-0 at the Go Summit in Wuzhen, China, from May 23 to 27, 2017. AlphaGo also defeated a Go team of five world champions on May 26, 2017 during this Go Summit.

We will never forget the experience of Ke Jie's confrontation with AI. He was full of confidence before the game, and he occasionally showed emotions at the scene, but he accepted the result calmly after the game, as if rehearsing a wonderful debut that AI was about to take a higher level.

After the tournament, DeepMind team announced that AlphaGo would not participate in future Go tournaments.

To this day, almost all of the world's top Go players have become accustomed to choosing AI training as the norm to improve their strength. Humans have easily accepted their former challenger as their "New Master". But beyond Go itself, can AI really make a peaceful transition from "challenger" to "New Master" for humans? Apparently, some people don't think so.

A week before AlphaGo's victory over Ke Jie, on April 27, 2017,

Stephen William Hawking, a famous British physicist, said in a speech at the Global Mobile Internet Conference in Beijing, "Humans must be alert to the threat of the development of AI... AI would take off on its own and re-design itself at an ever-increasing rate. Humans, who are limited by slow biological evolution, couldn't compete and would be superseded."[1]

Stephen Hawking was a staunch "AI threat" advocate, and until his death a year later, this remained one of his most important pieces of advice to humankind.

When the world of Go was meeting the head-on challenge of AI, the public's reaction was changing from curiosity, shock, trepidation, and then to calmness, which seems to be a microcosm of the collective reaction of the whole world to AI.

In the following five years, AI replaced more and more links of human work, but it never had the same impact as AlphaGo because more transcendence occurs outside the sight of the public.

Artificial Intelligence's First Year of Application

In 2017, almost everyone snapped to attention, with their ears full of news about AI. Huawei released its first AI chip, Alibaba invested 100 billion to set up the Alibaba DAMO Academy, and Baidu CEO Robin Li was seen sitting in a self-driving car in the livestream video on the fifth ring road in Beijing, thus earning the first ticket for domestic self-driving.

In the next year, Robin Li was ranked among Top 10 Global AI Figures by *Harvard Business Review* (Chinese Edition) the following year, the only Chinese business leader to be included on the rankings. Robin Li is at the forefront, as well as the architect of China's AI industry. He built the first self-driving open platform in the world, promoting the development of the smart driving sector in China and globally.

In October 2017, Sophia, a humanoid robot, was granted Saudi citi-

---

1   Wang Xinxin, *The Paper*, "Video | Hawking's Speech in Beijing: Artificial intelligence May Also Be the Terminator of Human Civilization", April 27, 2017.

zenship at the Future Investment Initiative in Saudi Arabia, which made her the first robot in history to do so. What caused a media frenzy was not the historic status symbol of the robot's citizenship, but the question posed by her creator, Hanson. He jokingly asked, "Do you want to destroy humans?... Please say no."

And Sophia's response was not quite what Hanson had in mind, "OK. I will destroy humans."[1] Sophia answered without hesitation. Whether this answer came from the joke of the creator or the real idea of AI, it was obviously enough to make everyone stunned.

In 2017, among a large number of new technologies, new hardware and new algorithms, AI was widely used. Compared with AlphaGo's challenge to human Go, in fact, in most scenarios, AI disdained to compete with humans, or even directly negated the meaning of the game as soon as it debuted.

In 2017 "Double 11" e-commerce carnival, Alibaba's AI designer "Luban" immediately took office, claiming that "1 second can make 8000 posters", which made a splash. None of the designers received the notice of the competition because no human had the qualification to participate in this level of competition.

Not all AI technologies had completed their own technical breakthroughs in 2017. However, in this year, the large-scale application of AI staged one unprecedented show after another for the public with super high frequency, super scenarios and super shocks.

2017, therefore, became the AI's first year of application.

Shock, sadness, excitement, and concern made up a rather complex emotions when humans faced AI in 2017.

Every country, every city, every enterprise and every individual clearly reached a critical moment that should be solemnly considered: "How to meet the arrival of the intelligent era? How to position oneself in the smart era?"

---

1 Wang Xiaoyi, *Shangyou News*, "Sophia, The First Ever 'Robot Citizen': I Will Destroy Humans", October 30, 2017.

Section II

Alternation Between Imagination and Practice, with Continuous Interaction for Five Years

One of the usual misunderstandings that needs to be dispelled when talking about the burgeoning AI is that, in essence, it is an advance in production technology, not a challenge to existing civilization. By recognizing this, many of the concerns about it may be tempered, or at the very least, may not influence the choice of action.

Globally, governments and innovative companies are far more optimistic and positive than physicists such as Hawking in the face of the coming age of intelligence.

A City's Answer to the Test of the Smart Era

In China, as early as March 2016, artificial intelligence was written into the 13th Five-Year Plan Outline; and by 2022, "artificial intelligence" has been written into the State Council's government work report for six consecutive years.

One of the most crucial points of time remains 2017. China put the development of AI on the agenda in a grand way from the national strategic level: on July 20, 2017, the State Council released the *Development Plan for the New-Generation Artificial Intelligence*, which puts forward the guiding ideology, strategic objectives, key tasks and safeguards for the development of a new generation of AI in China towards 2030, deploys the construction of a first-mover advantage in the development of AI in China, and accelerates the construction of an innovative country and a world power in science and technology.

How important was this development plan in 2017?

This issue needs to stand in a longer historical dimension to evaluate more objectively. However, this does not affect us in any way today, five years later, to feel the powerful impetus of the high-level national strategy for the intelligent industry and the profound changes brought about by it.

It can even give rise to an afterthought of how passive the country's development planning would be if it came a little late, even if only a year late.

From 2016 to 2017, the first AI strategies were released by the United States, China, Japan, and the UK, respectively. From 2017 to 2018, mainstream powers around the world were leaping into action with an eye on a strategic competitive layout for the next era, the core was the AI industry. By 2018, almost all major economies around the world, such as Germany, France, Italy, India, South Korea, Russia, and Singapore, had launched their own AI development strategies one after another.

Similarly, it seems to be the first time in history that almost all countries in the world collectively released their own country's industrial development strategies in close proximity to the point of time. Even when it came to the 5G strategy involving the core national information infrastructure, the strategic planning actions of each country were not so neat.

A race among countries around AI was launched in 2017.

Five years later, the global AI market is projected to grow by almost a fifth to reach $432.8 billion in 2022, according to a new report from IDC. The analyst firm expects spending on AI to increase by 19.6% this year, which includes spending across hardware, software, and services.[1]

Back in China in 2017, under the stimulating effects of the development plan of the State Council, the investment and financing around AI in the corporate and VC circles showed an unprecedented boom. Qichacha shows that the number of financing events in the AI track from 2016 to 2018 continued to remain above 900, and the market had some rational return after 2019, but was still above 500.

Under the guidance of the development plan, different provinces

1 Shiyi, *People's Posts and Telecommunications News*, "IDC: Global AI Market Size Reaches $432.8 billion in 2022 with Nearly 20% Growth", March 9, 2022.

and cities in China have also started the AI development race and have started the in-depth layout of the intelligent industry from their own perspectives.

However, the intelligent industry is a huge industry involving almost all the economy, life and public governance of the whole society, and the industrial base, social resources and talent composition of each province and city are completely different. What kind of attitude, from which angle to embrace the smart industry, has become a time-limited must-answer in front of each province and city.

If we can offer a satisfying answer, we will seize the opportunity of the times and change the fate of the city; if not, we may miss the opportunity and will face the future passively.

After the release of the *New Generation Artificial Intelligence Development Plan,* Fuzhou, Guiyang, Chongqing and Shanghai, which responded the fastest, took "digitalization", "big data", "smart industry" and "artificial intelligence" as the keywords for the first time. In 2018, they applied to hold top summits: Digital China Summit, The International Big Data Industry Expo, Smart China Expo and World Artificial Intelligence Conference.

After that, summits, exhibitions, forums and fairs themed on the keywords of AI, digitalization, big data, cloud computing and other smart industries have sprung up in major cities across the country, while the smart industry summits, which are jointly sponsored and supported by multiple national ministries and commissions at the top international level, are only available in the above four cities.

It is worth mentioning that, among the four major intelligent industry summits in 2018, three of them were applied for and held for the first time, while "2018 The International Big Data Industry Expo" held in Guiyang was actually the second one. As early as in July 2017, two months before the State Council's *New Generation of Artificial Intelligence Development Plan* was issued, Guiyang successfully held the first "Big Data Expo", which showed a keen sense of foresight for the coming of a big era.

In any case, the inevitability of the times and the serendipity of the city together determine the key answer of Chongqing, the city of manufacturing, to the future intelligent industry.

Four Sessions of Smart China Expo, Five Years of Transformation

Every year in July and August, the summer in Chongqing continues to be scorching. Chongqing citizens used to go out to escape the heat, but in recent years, many people have adjusted this living habit.

Since 2018, Smart China Expo has been permanently settled in Chongqing. Every August, Smart China Expo is held on time, which has become a new summer expectation and new habit for many people.

Intelligent industry scientists, entrepreneurs, innovators and researchers from all over the world gathered in hot Chongqing to attend the hot Smart China Expo. Focusing on AI, digital economy, intelligent manufacturing and other topics, they have delivered the most cutting-edge innovative theories, innovative knowledge, innovative skills, innovative experiences and innovative models to the world, clashing with new wisdom and contributing new achievements.

Since it is the most advanced, of course, it must be the hottest. For Chongqing, it is the original temperature of the furnace city, and also the future vision of a smart city.

How important can an event that lasts only a few days be to a city? Perhaps many readers will ask a similar question.

Let's temporarily put aside the various guest speakers, forum dialogues, product demonstrations, interactive screens, and first look at a set of data.

In 2018, 501 major projects with a total investment amount of 612 billion yuan were signed at the Smart China Expo.

In 2019, 530 major projects with a total investment amount of 816.9 billion yuan were signed at the Smart China Expo.

In 2020, affected by the epidemic, the Smart China Expo was mainly held online, and many enterprises and project parties could not come to the site, even so, 71 major projects were signed, with a total investment amount of 271.2 billion yuan.

In 2021, 92 major projects with a total investment amount of 252.4 billion yuan were signed at the Smart China Expo, although the overall scale was comparable to last year, but the projects with a total investment more than 1 billion yuan and 5 billion yuan had increased signifi-

cantly, and the volume of a single project had become larger.

This running account made the official read very sonorously, the media reported it proudly, and the citizens were very excited to see it.

The exchange of ideas and the signing of projects are linked together, which is a double spark.

The exchange of ideas has opened Chongqing's brain to understand the world and to take a closer look at the changing intelligent era; and the signing of the project has opened Chongqing's gates to welcome guests and to join hands to create a better intelligent industry in Chongqing.

With the help of the Smart China Expo, as a popularity city in western China, Chongqing is rapidly rising both domestically and internationally. For global quality enterprises and outstanding talents, Chongqing has timely released the vitality of the city of innovation in the era of intelligence, attracting them to Chongqing to start their business.

As a landing project of the 2019 Smart China Expo, Chongqing Industrial Big Data Innovation Center has helped a number of Chongqing enterprises achieve costs reduction and improvement by promoting the deep integration of information technology and real economy, such as Internet and big data in the past two years. At present, the innovation center has built 8 intelligent transformation demonstration factories in the city and will provide "cloud" services for more than 7,000 SMEs in the year.

On August 5, 2021, Chongqing Economic and Information Commission revealed that the first half of the year, Chongqing added 16,000 enterprises "on the cloud, on the platform", and the city has more than 70,000 enterprises on the "cloud".[1]

In 2019, 2020 and 2021, a total of 125 enterprises were selected in the list of Chongqing technological innovation demonstration enterprises.

If we want to enumerate, there are many dimensions for the similar data; if we want to list, there are many similar breakthroughs, and many achievements. Behind these data, readers can feel that a city is revital-

---

1 Xia Yuan, *Chongqing Daily*, "Smart China Expo Landing Projects to Help Speed Up Industrial Internet Development, More Than 70,000 Chongqing Enterprises Go on the 'Cloud'", August 6, 2021.

ized in the intelligent era, thinking is changing, technology is iterating, industries are growing, good news is spreading frequently, and technological innovation breakthroughs of Chongqing enterprises are accelerating.

To attribute these figures retroactively, it is clear that the final answers will all point to the beginning and continuation of the Smart China Expo.

The Changing "Center Position" of the Big Stage of the Smart China Expo

From 2018 to 2022, the change of "Center position" on the stage of Smart China Expo also reflects a change of times and the evolution of a city.

At the 2018 Smart China Expo, in addition to government officers and academic experts from various countries, the speakers who stood on the stage and voiced for the arrival of a new era were all Internet entrepreneurs. They have been surfing in the digital world of the Internet for many years. They have a natural keen sense of smell for the surging wave of digitalization and intelligence, as well as the lofty sentiments of active embrace, and more of their speeches reflect their lofty aspirations for the future.

By 2021, the influence of digitalization and intelligence at the Smart China Expo is becoming stronger, and more and more entrepreneurs from traditional manufacturing industries are among the speakers. In their sharing, digitalization and intelligence are commonplace, and most of the content is no longer imagination and prophecy, but full of implemented technologies, practical solutions, achieved results and precipitated data.

In this case, the real economy has become a stage for artificial intelligence to "show its strength". In fact, in the past five years, most of the achievements of the intelligent industry have come from the extensive cooperation between traditional manufacturing enterprises and digital innovative platforms. Actually, it shows that China's Internet industry and traditional manufacturing industries have successfully completed

the connection and integration of digitalization and intelligence.

In order to deeply participate in international cooperation in digital economy and continuously expand foreign exchange and cooperation, the 2021 Smart China Expo and Shanghai Cooperation Organisation (SCO) Digital Economy Industry Forum were held at the same time. SCO Secretary General Norov was particularly impressed by the speed of development of China's digital economy, as he mentioned in his speech: *"The White Paper on the Development of China's Digital Economy (2021)* shows that China's digital economy is the second largest in the world, and the growth rate of China's digital economy is more than three times that of its GDP, which also shows that the digital economy plays a key role in driving economic development. In 2020, China's digital economy reached 39.2 trillion yuan (about $6 trillion) or 38.6 percent of GDP, effectively supporting epidemic prevention and control efforts and the country's economic development."[1]

In essence, the technology and concept of AI need to be converted into the productivity of digital economy, which is in line with the general trend of the era of intelligent industry.

The changes on the stage of the Smart China Expo, which is permanently settled in Chongqing, happen to be another microcosm of the economic structure of Chongqing and China as a whole.

---

1   Su Xiao, *People's Posts and Telecommunications News*, "China's Digital Economy Development White Paper Released, China's Digital Economy Reached 39.2 Trillion Yuan in 2020", May 6, 2021.

Section III

The Innovation of Ideas and Products Coincides with Smart China Expo for Five Consecutive Years

The successive holding of the Smart China Expo has largely changed Chongqing's way of interacting with this world. In the era of the rapidly changing intelligent industry, the city's development ideas are exceptionally clear and the entrepreneurs who have settled in Chongqing have shown great decisiveness.

Throughout the rise of famous cities in ancient and modern times, it is easy to find that, at a historical juncture between turbulent times, only when the top design of a city and the bottom struggle reach a key consensus, can the city be renewed in the future-oriented growth. When people of Chongqing clearly see the rising opportunities given by the smart era, they will forge ahead determinedly.

And for Chongqing, the holding of the Smart China Expo is not only to hold a successful meeting, but through the Smart China Expo to complete the reengineering of a city's smart industry gene. In order to upgrade manufacturing, we have to connect factory equipment; to improve life, we have to innovate intelligent life; to optimize governance, we have to open the gates to the street data. Once it is connected, innovated and opened, we could talk about the experience and the gains together. When we review the precipitation, look forward, and start again, the scenery is completely different.

Xiantao Big Data Valley, from a Barren Mountain to a Hot Destination

In April 2014, Chongqing announced a brand-new plan for a sci-

ence and technology industrial park. The total planning area is 2,674 acres, and the site is located in the village of Xiantao, Shuanglonghu Street, Yubei District, Chongqing. As this location is mentioned, Chongqing citizens generally lament that "too far away". Just how far off is it?

At that time, in the minds of Chongqing citizens, the recently completed Chongqing Garden Expo is far enough from the main city, and the place is still 6 km due north of the Garden Expo.

At the beginning of the planning, the place was still a barren mountain slope, overgrown with weeds and rocks, without a decent road, without a clear sign, without perfect electric facilities, and without a unified water supply and drainage system.

The gradually completed Chongqing Xiantao Big Data Valley, with the accelerated construction of the "Innovation Valley", "Wisdom Valley" and "Ecological Valley", has now gathered thousands of enterprises, more than 5,000 all kinds of scientific and technological innovation talents, developed into a big data industry ecological valley, won many awards such as "China's most dynamic software park" [1].

Riding the waves of the intelligent era, the transformation of this barren mountain to a hot destination took only eight years.

How popular is Xiantao Big Data Valley?

It has become the preferred destination for the second headquarters of intelligent vehicle software enterprises nationwide. Chongqing Automotive Software Industry Base is officially settled here and has gathered more than ten key enterprises in the smart car industry, such as Chang'an Software, ThunderSoft, Thundercomm, and Black Sesame Technologies, with nearly 1,000 relevant practitioners. The first 5G self-driving open road scenario demonstration operation base has been built, which can conduct demonstration operation of intelligent network-connected vehicles on a 5-km-long circular route.

After eight years, there are a unique "Little Barbarian Waist", the 100-meter high corridor "Lord of the Rings", the "all-glass cube" conference center, the building block-shaped Big Data Academy, and also other buildings with a strong sense of technology.

---

1 Zhang Yizhu, *Chongqing Daily*, "Xiantao Big Data Valley, Where the Charm Is Big", August 5, 2021.

As of March 2022, Xiantao Big Data Valley had registered 1,293 enterprises, 250 enterprises, applied for 1,749 intellectual property rights, authorized 1,232 patents, and achieved steady growth in the amount of copyright registration, patent softwares and other intellectual property achievements.[1]

The Big Data Valley has now become a hot and competing destination for the global smart industry.

From a barren to a hot destination, the eight-year transformation of Xiantao Big Data Valley is precisely a microcosm of Chongqing's intelligent industry.

Smart China Expo, Chongqing's Second Investment Promotion Bureau

Xiantao Big Data Valley is just a microcosm of the city of Chongqing into the era of intelligence. In the past five years, the transformation similar to Xiantao Data Valley has actually happened simultaneously in every corner of the city.

Perhaps initially, for many guests attending the 2018 Smart China Expo, it was just an ordinary summit. Flying to Chongqing, attending the event, giving a speech, and then it's time to hit the road back to where their businesses are and continue to explore the future of the smart era.

Later, attending the Smart China Expo becomes a door for many people to enter Chongqing. For Chongqing, the Smart China Expo has also become the "second investment promotion bureau".

In 2018, because of the Smart China Expo, leading domestic technology companies such as Tencent, Alibaba, Huawei, Baidu, Inspur, China Electronics Corporation and iFLYTEK have landed one after another. They either set up their headquarters or built customer experience and data centers of southwest China in Chongqing and started to explore AI more deeply in Chongqing.

---

1   Liang Haonan, *CQNEWS*, "Chongqing Unique! Xiantao Big Data Valley Selected as National Copyright Demonstration Park", March 19, 2022.

In the following three years, countless enterprises have entered Chongqing's intelligent industry ecosystem through the stage of the Smart China Expo, especially in the fields of intelligence, big data, high-end electronic information and new materials.

On August 23, 2021, the news that Li Shufu appeared in Chongqing and attended the Smart China Expo quickly made the headlines on the major media in China. What kind of in-depth cooperation would this legendary entrepreneur in the automotive industry have with Chongqing, an automobile city known as "Detroit in the East"? It has become a sensational topic in the automotive industry.

For a long time, the public image of this Zhejiang entrepreneur was inextricably linked to standard traditional Chinese manufacturing. Geely has continued to expand its footprint around auto manufacturing and is best known for its ambitious feat of acquiring Volvo. Along with the rise of Chinese manufacturing in the world, Li Shufu has become one of the representative figures of the Chinese manufacturing industry.

What is less known, however, is that while a series of industrial layouts have been completed, Li Shufu and Geely Automobile have been equally aggressive in the layout of intelligent manufacturing.

Li Shufu of the next era started from Chongqing. Moreover, it was no longer a sensational news that Li Shufu came to Chongqing.

Within a year, Geely made several important layouts in Chongqing. First, Geely Group participated in the reorganization of Lifan Motor, and then took a stake in Lifan. In just half a year, Lifan quickly turned losses into profits, and officially changed its name to "Lifan Technology". Secondly, Polestar, a strategic high-end new energy brand of Geely Group, was established and put into production in Chongqing. The third big move was made public through the 2021 Smart China Expo, that is, GYMD Digital Technology, the global headquarters of Geely Group's industrial Internet, settled in Chongqing.

On the first opening day of the 2021 Smart China Expo, in the main exhibition hall, facing the entrance was the digital twin booth of GYMD's "New Model of Intelligent and Flexible Manufacturing"; behind the booth was Geely's hatchback intelligent concept car; and in the main venue of the opening ceremony of the 2021 Smart China Expo, Li Shufu was giving a speech as a special guest. In his speech, he said,

"Chongqing is a new highland for the development of China's digital economy, and also an important modern manufacturing base with a very long history of industrial development. It not only gathers a large number of advanced manufacturing enterprises, but also has the policies and basic capabilities to support the rapid development of industrial Internet and digital economy, which is a fertile land with great development potential."[1]

Obviously, he regards Chongqing as his "home".

1 Li Hui, *CQNEWS*, "Geely Holding Group Chairman Li Shufu: Chongqing Is a Fertile Land with Great Development Potential", August 23, 2021.

Section IV

Reviewing the Five-Year Smart China Expo, the Development Opportunities Bestowed by the Times to Chongqing

It is difficult to find an absolute starting point for this great era of intelligent industry. Both the accumulation of technology and the promotion of application are generated step by step, just as little strokes fell great oaks.

However, when we look back on the way of the smart industry, we can always find some special moments. Various characters came to stage, various technologies spurt out, various views were inspiring, and various policies were made at a high level. Look at any single point, it seems to be accidental. But when there are too many accidents in a period of time, it precisely announces the inevitability of a brand-new era making its debut.

From 2018 to 2022, for AI , the five years is the journey of countless chances gathered into inevitability. The genesis and continuation, sprouting and growth, ascension and precipitation, integration and harvest of the Smart China Expo held in Chongqing all happened during this period, with the times and with the world.

It can be said that through the five-year Smart China Expo, we can witness the whole world from one era to another, and even witness Chongqing, from influencing the world to changing ourselves.

A Critical Moment for a City

For those who are familiar with regional economies, it is easy to see that different cities have very different styles of products and cus-

toms, and also have very different economic characteristics. Why are some cities good at industrial manufacturing, some good at agricultural farming, some supplying clothing to the world, and some with delicious food to the whole country?

What determines the unique industrial characteristics of a city standing out across the country and around the world?

Before the global economic integration, this question is actually not difficult to answer. Backer has mountain, draft relying on water, which is an objective law summed up by human ancestors. However, when the whole world's economy is integrated and connected, the scope of industrial division of labor is constantly expanding from one street to one village, one town, one county, one city, one province and one country, and finally to the whole world. Many conditions, such as climate advantages, geographical locations, property resources and so on, which originally determined the characteristics of urban industry, gradually become invalid.

As the saying goes, "Thirty percent destiny and seventy percent diligence." The factors that dominate the city's destiny have changed from relying on natural resources to determining the path to seize opportunities.

At first, in the process of urbanization, some industrial clusters were formed by spontaneous experiments of market economy. Of course, this process will be full of contingency. A self-employed person who becomes rich first or an accidental successful enterprise will bring a strong demonstration effect to the local area, thus triggering industrial clusters. From the vertical point of view, successful enterprises often drive the improvement of the industrial chain in a market-oriented way through upstream and downstream matching. From the horizontal point of view, successful enterprises tend to produce a model effect locally, which can inspire more and more local enterprises, thus forming a larger industrial scale.

This kind of industrial development by point with line and line with scale is often the more typical logic for a city to form its own unique economic characteristics in the process of urbanization.

However, at a critical juncture in the history of revolutionary innovation in human society, it is clearly impractical to rely solely on the purely spontaneous evolutionary capacity of a city to drive a city

through rapid industrial upgrading or transformation.

This is a very test of the wisdom of city managers, how to look to the future, grasp trends, develop imagination and build a course of action to guide the transformation of a city and change its destiny.

Through high level planning, a city relies on its own industrial base to embrace trends and complete the overall upgrading of urban industries, with numerous successful cases in the forty years of reform and opening up.

In the 1990s, Shanghai seized the opportunity of China's financial market recovery to become a financial city that has stood out globally.

In the 2000s, Shenzhen seized the opportunity to become a city of technology that influenced the world as China has become a powerhouse of science and technology innovation.

In the 2010s, Hangzhou seized the opportunity of China's e-commerce explosion and has become a leading global e-commerce city.

Facing the third decade of the 21st century and the new opportunities for the development of global intelligent industries, Chongqing has a clear goal.

Over the past 100 years, Chongqing, a manufacturing center in the west, has been playing the role of a manufacturing base: from saving the nation by engaging in industry and building factories in the early 20th century to the westward relocation of industry, from the third-front construction of heavy industry after the founding of the People's Republic of China to the rise of light industry in 1980s, from the military industry to civilian use, and the automobile and motorcycle industry standing out in the 1990s to the the rapid growth of automobile manufacturing in the 21st century.

After China's entry into WTO, factories in coastal cities have mushroomed, and the manufacturing industry in western China develops slower than that in eastern China. Even so, Chongqing has not lost its position as a manufacturing center.

With the industrial adjustment of coastal manufacturing industry, when precision electronic manufacturing such as notebook computers and mobile phones began to look for new destinations in western China, Chongqing once again seized the opportunity. It has undertaken the transfer of coastal manufacturing industry and introduced many high-

end manufacturing enterprises such as Hewlett-Packard, Foxconn, Quanta, Inventec, Pegatron, Wistron and Compal. The perfection of the whole category manufacturing territory has injected modern genes into this traditional manufacturing city.

It seems to be waiting for the arrival of a real explosive moment. Chongqing will embrace intelligent manufacturing and stand in the "Center-position" of a huge industry facing the future.

The development of the AI industry has allowed the city to discover the arrival of a greater era of its own.

On July 4, 2022, the AITO M7 intelligent car was officially launched, with orders breaking 20,000 units in 2 hours of pre-sale and 40,000 units in 4 hours of pre-sale, which is another result of the in-depth cooperation between SERES Auto of Xiaokang Group and Huawei in cross-border car manufacturing.

Looking back at the entire AITO, the M5 released a year ago by both partners has an equally bright performance. Since the M5 started delivery in March 2022, Xiaokang shares ushered in an explosive turning point, advancing into the top five of the high-end new energy SUV sales list that month, completing 3,000 units in the first month of delivery, and subsequently maintaining the top ten of the new energy SUV sales list, delivering 11,296 units in a total of 87 days, setting a new record for the fastest delivery of a single model of the brand to break 10,000 units.[1]

The cooperation between Xiaokang and Huawei has pioneered a deep cross-border integration of the joint business of intelligent vehicles. Both partners deepen the cooperation in promoting new energy vehicles, jointly create high-performance and intelligent mobility solutions, provide users with more efficient and convenient intelligent vehicle products and intelligent mobility experience, and bring the digital world into every vehicle.

Relying on Chongqing's high level layout in the intelligent industry and the linking opportunities that the Smart Expo continues to create for local manufacturing enterprises and the global intelligent industry, Hua-

---

1　Yan Wei, *Shangyou News*, " Production and Sales Data Released in May, Chongqing Auto Market Glows with 'New Green' ", June 2, 2022.

wei's cooperation with Xiaokang is not isolated. More and more Chongqing manufacturing is rapidly becoming the pioneer of innovation in the intelligent era, and behind the countless cases of intelligent and digital transformation of enterprises, a more ambitious urban development strategy of making a city of science is being carried out in parallel.

On January 3, 2020, General Secretary Xi Jinping, at the sixth meeting of the Central Finance and Economics Commission, made special arrangements for the construction of the twin-city economic circle in the Chengdu-Chongqing region, and put forward clear requirements for the promotion of science and technology innovation in the Chengdu-Chongqing region.[1]

Subsequently, in the Chongqing High-Tech Zone, the construction of the Western China (Chongqing) Science City opened a high-speed mode: within 3 months, Chongqing held a meeting of the Municipal Planning Commission and the City Urban Enhancement Leading Group, and the meeting considered the *Western China (Chongqing) Science City Territorial Spatial Planning (2020—2035)*; within 9 months, the mobilization meeting for the construction of the Western China (Chongqing) Science City was successfully held, focusing on the construction of 79 key projects such as Science Avenue, Science Valley, and the demonstration project of ecological water system of Science City were started, marking the full-scale construction of the Western China (Chongqing) Science City; within 10 months, Chongqing High-tech Zone held a centralized signing activity of key innovation platform projects of universities and research institutes in the Western China (Chongqing) Science City, and a total of 24 projects were signed in this signing activity.

Since the Western China (Chongqing) Science City started construction, based on achieving a high level of scientific and technological self-sustainability, a large number of large devices, large institutions, large platforms, large projects continue to accelerate: China's Natural Population of Biological Resources has recruited sample collection of more than 100,000; 1 Digital Transformation Center of Peking Uni-

---

1 *Xinhuanet*, "Xi Jinping Chairs the Sixth Meeting of the Central Financial and Economic Commission", January 3, 2020.

versity Chongqing Institute of Big Data has been established, and 13 big data intelligence-related frontier laboratories are accelerating scientific research; Chinese Academy of Sciences Automotive Software Innovation Platform landing construction has been started and the core research team has been stationed; Shanghai Jiaotong University Chongqing Artificial Intelligence Research Institute has been officially established ...[1]

In the third decade of the 21st century, the Western China (Chongqing) Science City is seizing every opportunity to accelerate the construction of a core area with national influence as a science and technology innovation center.

From a narrower perspective, AI is technology. Whether to speech recognition, visual recognition, deep computing, neural computing, edge computing, or big data, strong technology genes are needed.

But from a broader perspective, whether the intelligent industry can form a competitive gathering in the smart era, the core is the industrial base. After all, technology can transfer and flow, while the industry needs more aggregation and rooting.

In terms of the technology-to-industry application route, it is far faster and more efficient to put advanced technology into manufacturing improvements than to incorporate the technology into the final product.

Therefore, Chongqing is an excellent destination and testing ground for the combination of AI and manufacturing.

1　Liu Hanshu, *Shangyou News*, "Accelerating the Construction of the Core Area of the Science and Technology Innovation Center with National Influence, the Western China (Chongqing) Science City Jinfeng Laboratory Is Put Into Operation This Month", May 12, 2022.

Part II

Intelligent Economy: A Standard Answer to Thousands of Industrial Problems

Every industry will encounter similar problems when it reaches its ultimate development; every era will encounter similar bottlenecks when it reaches its destination; and whenever the economy and society have accumulated sufficient problems and encountered sufficient stubborn bottlenecks, the iteration of existing technologies has reached its extreme, and the buzz of past experiences is suddenly silent. At this time, there is always a new industrial revolution coming out of the blue.

Thousands of industrial problems in the economy and society, such as the contradiction between large-scale production and customization, the rift between the physical world and the virtual world, the gap between the real economy and the digital economy, the difference between the digital consciousness and the social algorithm, finally successfully obtained a new standard answer from the times—intelligent economy.

Section I

Solving Industrial Problems Is the Consensus of Global Intelligent Industry Development

From August 23rd to 25th, 2018, Chongqing International Expo Center, the first Smart China Expo was being held in full swing. Chen Wenguo, a 65-year-old citizen of Nan'an District, Chongqing, who has been paying attention to the development of autonomous driving technology, finally saw a real self-driving car.

In addition to self-driving cars, vending machines with face ID payment, VR simulated diving to underwater museum, 3D printing, drone flight, etc. also amazed Chen Wenguo. According to statistics, there were more than 500,000[1] spectators at Smart China Expo, and they were as amazed at the "black technology" in the intelligent age as he was. In response to people's enthusiasm, the organizing committee even made a decision to extend it for one day.

If you travel through time and space, you will find that similar scenes were staged in Hanover, Germany, which is nearly 8,000 kilometers away from Chongqing.

Back to more than 70 years ago, from August 18 to September 7, 1947, the first Hannover Messe was held in Hannover, Germany. People were attracted by the world's smallest diesel engines, dentures, foldable baby strollers and Beetle cars at the exhibition at that time and imagined the great impact they would have on human life.

Hannover Messe has been held once a year since 1947, and it has become a platform for international technology and industry exchange. Just as Smart China Expo symbolizes that the era of intelligent industry

---

1 Yang Ye, *Shangyou News*, "500,000+! This Number Stunned Alibaba and Tencent", August 26, 2018.

is accelerating, Hannover Messe also has its significance of marking the times. It is the product of the times of electrification and automation. It was also at Hannover Messe in 2013 that the concept of "Industry 4.0", the Fourth Industrial Revolution, was officially launched, and the curtain of the era of intelligent industry was officially opened.

Industry Consensus with Different Expressions

The concept of "Industry 4.0" was born in 2011, and its proponents were Henning Kagman, former president of the German National Academy of Science and Engineering, and Wolfgang Walst, former director of the German Artificial Intelligence Research Center.

How to understand "Industry 4.0"?

Henning Kagermann, one of the proponents of the concept, thinks: "It needs to be understood from two aspects. On the one hand, everything is interconnected, that is, the Internet can be related to many links in life; on the other hand, the integration of the physical world and the virtual world, that is, changes in the virtual level can immediately affect the physical level, and vice versa."[1]

"Industry 4.0" has a more specific interpretation that is closer to production. Andreas Pinkwart, dean of Leipzig Graduate School of Management, thinks: "Networking, digitalization and intelligence are the means to realize 'Industry 4.0' rather than the goal. The fundamental goal is to change the existing thinking mode of production, from mass production to small-batch customized production oriented to customer demand, to complete personalized consumption with high efficiency and high quality, and to control costs. The task of 'Industry 4.0' is not simply to upgrade the digital network of industrial production, but to integrate products, consumption and production into a network that can communicate with each other. "[2]

---

1 Feng Xuejun, Guan Kejiang, *People's Daily*, "Hannover Messe Presenting 'Digital World'", April 15, 2015.
2 Feng Xuejun, Guan Kejiang, *People's Daily*, "Hannover Messe Presenting 'Digital World'", April 15, 2015.

The above contents are all early interpretations of "Industry 4.0" by experts. Different perspectives lead to different interpretations. With the passage of time and the development of new technologies, the connotation of "Industry 4.0" is changing and enriching. However, one thing will not change, which is reflected in the title of the report *Securing the Future of German Manufacturing Industry: Recommendations for Implementing the Strategic Initiative INDUSTRIE 4.0*, submitted by the German "Industry 4.0" working group to the German federal government in 2013. Just as the title of the report indicates, "Industry 4.0" represents the "future of manufacturing industry", and the means of production, production methods, technological innovations and business ideas that map the future of manufacturing industry are all within the interpretation scope of "Industry 4.0".

"The future of manufacturing" is expressed as "Industry 4.0" in Germany, while in other countries, there are different expressions. The United States calls it "industrial Internet", Japan calls it "interconnected industry", and China mainly uses "intelligent manufacturing". These are the expressions of different countries on the future of their respective manufacturing industries. In fact, all countries have the same demand for new technologies and infrastructure such as big data, AI and 5G, but they also have their own priorities based on different economies, industrial bases and industrial concepts.

There is a consensus that "Industry 4.0" in Germany focuses on the intelligent upgrading and intelligent standards of equipment, which is consistent with the original industrial path of "powerful equipment" and equipment output in German industry. China's intelligent manufacturing emphasizes the deep integration of information technology and manufacturing industry. Comparatively speaking, China has a broader perspective in the industrial chain, involving not only the production, but also the information integration between the consumption and the production. The foothold is to improve the overall level of manufacturing industry, which lies in the quality, efficiency and costs of product production, behind which is the basic status of China as a manufacturing power.

No matter how it is expressed, no matter what the emphasis is, it has become the consensus of major countries in the world to develop intelligent industries based on new technologies and concepts.

The Driving Force of Intelligent Industry

Ten years after "Industry 4.0" was put forward, the intelligent industry has gone into the fast lane of vigorous development.

Why has the development of intelligent industry become a global consensus? What is its driving force? Why is it as irresistible as the historical trend?

Of course, the reasons are complicated and hierarchical. It may be a superficial market need or a deep competition drive. It may be the natural germination of mature objective conditions, or it may be the future desire of subjective conscious thinking.

The consensus on the intelligent industry is firstly the consensus on the manufacturing industry. Although more than 10 years had passed since the international financial crisis in 2008, its impact is still playing a role, and it has a far-reaching impact not only on the financial sector, but also on the manufacturing industry. Some people think that the hollowing out of manufacturing industry and the dominant position of virtual economy are not conducive to economic stability, but also the culprit of high unemployment rate.

One year later, in November 2009, former US President Barack Obama proposed that the U.S. economy should transform from the high consumption pattern of financial credit, and rebalance the manufacturing industry and the service industry. The trend of economic policy thoughts of "from theory to practice" and "moving forward with the combination of theory and practice" has brought the manufacturing industry back to the spotlight once again, and continues to dominate the subsequent global economic competition pattern.

On one hand, relying on manufacturing industry to resist economic risks is the ballast of social economy; on the other hand, the development of intelligent industry can create huge economic output value by itself. AskCI Resarch predicts that the market size of China's intelligent manufacturing equipment will exceed 2.6 trillion yuan in 2022,[1] which

---

1   *The Economic Observer*, "Human Eyes Become 'Smart Eyes' in Traditional Quality Inspection , Lenovo Algorithm Empowers Enterprises to Transform and Win Again", December 29, 2021.

is a rapid growth compared with the market size of less than 1.8 trillion yuan in 2019.[1]

How much impact can Germany's "Industry 4.0" have on Germany's economic output? It is estimated that in 2025, Germany's output value in six industries of industrial manufacturing, automobile, chemistry, IT, electronics and agriculture will increase by 78.7 billion euros, and the overall output value of Germany will increase by 267 billion euros.[2] This output added value alone is not surprising, but when compared with the total GDP of Germany of 353.9 billion euros in 2021, it is not difficult to find that it has brought a large increase in the proportion of the German economy.

In addition to the promotion of economy and the driving force to development of intelligent industry, it also includes the competition and cooperation between countries.

As early as in 2014, the international scientific community interpreted the concept of "Industry 4.0": "The name 'Industry 4.0' means the Fourth Industrial Revolution in human history."[3] Also in 2014, in the *Sino-German Cooperation Action Plan* signed by both parties, the relationship between "Industry 4.0" and "Intelligent Era" was clearly defined: "Industry 1.0 is the era of steam engine, Industry 2.0 is the era of electrification, Industry 3.0 is the era of information, and Industry 4.0 is the era of using information technology to promote industrial change, which is also the era of intelligence."[4] The so-called "Fourth Industrial Revolution" is a brand-new technological revolution based on AI, clean energy, automatic control technology, quantum information technology and virtual reality. Just as the age of steam is to Britain, the age of electricity is to Germany, and the age of information technology is to the

1　*Qingdao Financial Daily*, "China Opens a New Round of Intelligent Manufacturing Construction Curtain", April 23, 2021.

2　Wang Luohan, Wang Weinan, *Outlook of Global Science and Technology Economy* (Vol. 36), "Review of the Development of German Industry 4.0, What Can China Learn from?", November 12, 2021.

3　Gao Yedun, *People's Daily Online*, "Definition of Industry 4.0: Germany Wants to Start the Fourth Industrial Revolution", January 13, 2014.

4　Li Songtao, *China Youth Daily*, "Li Keqiang's Visit to 'Industry 4.0' in Europe Is Timely for China", October 14, 2014.

United States, the energy it carries will inevitably lead to changes in the world's competition and cooperation patterns. No one is willing to miss this round of industrial wave, and no one is willing to lose its position in the industrial chain in the current globalization.

If we focus our attention a little more, then we may find that the intelligent industry composed of intelligence and manufacturing can solve many practical and specific problems.

In a statement made by Henning Kagmann, the father of the concept of "Industry 4.0", these problems are concentrated: the idea of "Industry 4.0" is that production is highly flexible, products are personalized, resources are saved, and older workers can get help from intelligent auxiliary systems. Production has become humanized, and workers can work at home or nearby. Intelligent factory is not an unmanned factory, it just broadens the space for workers to participate and make decisions. The number of people working directly in the first line of the workshop is reduced, but others can participate in the work of system scheme, research and development and coordination.

"Production flexibility, product personalization" can solve the problem of overcapacity in the industry. Uncontrolled production regardless of the real market demand is the main cause of overcapacity. Intelligent production mode realizes the interconnection between people-to-equipment, equipment-to-equipment, and extends the tentacles of data collection and sharing to the demand, thus realizing more refined production.

The high degree of automation and self-decision ability of intelligent industrial equipment will reduce the dependence on human labor in the production process and change the time and space for workers to participate in manufacturing. On the one hand, it can cope with the rising labor costs and reduce the costs; on the other hand, it can also cope with the decreasing trend of the employed population in all countries around the world.

At the same time, we can't ignore that every industrial change is closely related to energy. Energy changes the way of production, and the way of production in turn seeks to change the type of energy. In the age of steam and electricity, the main support of energy was non-renewable resources such as coal and oil, which were not friendly to the environment. Now this support is weakened, but it is still the main power

source of the whole human society.

How does the intelligent industry solve the energy problem? One keyword is saving, and the other keyword is substitution. When the production is more accurate, the interconnection is smoother, and the energy is naturally saved. As for the replacement of energy, we can see that photovoltaic, new energy vehicles and other industries will reduce the dependence of human beings on non-renewable energy in the intelligent age.

Rome was not built in a day, and the door of intelligent industry was not suddenly pushed open in a certain year. The development of industry sometimes does not completely follow the will of human beings, and it has a spontaneity, especially when the technical elements are fully available. The development and maturity of many new technologies, such as robots, Internet, Internet of Things (IoT), big data, cloud computing, etc., are the strengths to open the door to the intelligent age and the puzzle to build the intelligent age.

The development of intelligent industry, superimposed with subjective and objective factors with rich levels, has a sense of fatalism, like the historical trend, which is unstoppable.

Section II

Laying Out an Intelligent Strategy, China Switches to the Dividend of the Times

If the proposal of "Industry 4.0" led to the intelligent age conceptually for the first time, when did the substantial germination of the intelligent age take place?

The year of 2008 may be an easy answer to reach a consensus.

2008 was destined to be an extraordinary year, with many important events, such as the global economic crisis and the Beijing Olympic Games. In fact, in 2008, there were three kinds of transcendence pointing to the opening of an era.

In the keynote speech at the opening ceremony of the 2021 Smart China Expo, Wu Manqing, general manager of China Electronics Technology Group Co., Ltd. and academician of Chinese Academy of Engineering, mentioned three simultaneous transcendences in the world in 2008: "First, the urban population exceeded the rural population, and urban population transcendence implies a human agglomeration; second, the number of machine connections exceeded the number of human links; and third, the number of mobile links exceeded the number of fixed links."

What is the layout of China in the lively digital world and the intelligent era full of imagination? If layout is a subjective wish, what are the objective conditions for developing intelligent industry? The objective condition of superiority is actually what we often call "dividend". What is China's "dividend of the times" for developing intelligent industries?

Intelligent Industrial Strategic Layout

China, the United States and Germany play an important role in the world economy and manufacturing industry. They have the consensus of developing intelligent industry and are the most advanced and complete in the layout and practice of intelligent industry strategy.

Each country's intelligent industry strategy must be put forward based on its own actual situation. Different strategic routes complement and refer to each other. By combing the strategic layout of intelligent industry in the United States and Germany, we may be able to understand China's intelligent industry strategy more clearly.

The United States' smart industry layout began with the return of manufacturing industry. In December 2009 and August 2010, Obama successively signed *A Framework for Revitalizing American Manufacturing and U.S. Manufacturing Enhancement Act of 2010*. The key words of the two acts are manufacturing industry, which reveal an obvious truth that the intelligent industry is manufacturing plus intelligence. If the manufacturing industry is hollowed out, the intelligent industry will become rootless weeds.

Later, the United States began to lay out "intelligent manufacturing", for example, *Network Charter Manufacturing USA Program* was signed in the form of an act, aiming at establishing research institutes in key fields to gather many organizations from industries, academia, federal and local governments, and establish and improve the innovation ecosystem. *A National Strategic Plan for Advanced Manufacturing* aims to connect small-and-medium-sized enterprises to participate in intelligent manufacturing and create an ecosystem. The key word of the two acts is "innovation". The former one is to stimulate innovation, and the latter one is to bring innovation to the ground and empower SMEs.

In June 2021, the U.S. Senate passed *The United States Innovation and Competition Act of 2021*, arguing that the U.S. federal government should strengthen the strength of new technologies in the United States through public investment in key areas, including chips, 5G networks, quantum computing and information systems. The key words of layout here are "key fields" investment, that is, the key breakthroughs in the development of intelligent industry.

Germany first proposed the concept of "Industry 4.0". In 2013, Germany officially released *Securing the Future of German Manufacturing Industry: Recommendations for Implementing the Strategic Initiative INDUSTRIE 4.0*, which raised "Industry 4.0" to the national strategic level.

The key word frequently mentioned in German intelligence industry layout is "digitalization". In 2016, Germany released *Digital Strategy 2025*, which developed 12 contents such as digital technology, reliable cloud, German data service platform, digitalization of small-and-medium-sized enterprises, and digitalization. In *The National Industrial Strategy 2030* released in November 2019, in addition to improving the framework conditions of industrial bases, strengthening the research and development of new technologies and mobilizing private capital, Germany believes that the most important breakthrough innovation at present is still accelerating the digitalization process.

The strategic layout of China's intelligent industry can be traced back to the major consulting project of *Research on the Manufacturing Power Strategy* initiated by Chinese Academy of Engineering and the Ministry of Industry and Information Technology in 2013, and the ten "key areas" later put forward as the key to the layout of China's intelligent industry, including: new generation information technology industry, high-grade CNC machine tools and robots, aerospace equipment, offshore engineering equipment and high-tech ships, advanced rail transit equipment, energy-saving and new energy vehicles, power equipment, agricultural machinery and equipment, new materials, biomedicine and high-performance medical devices.

On December 21, 2021, the Ministry of Industry and Information Technology, together with the National Development and Reform Commission, the Ministry of Education, the Ministry of Science and Technology and other departments, issued the *Smart Manufacturing Development Plan for the 14th Five-Year Plan*, which put forward three main goals for 2025.

The three main objectives include:

First, transformation and upgrading have achieved remarkable results. More than 70% of manufacturing enterprises have basically realized digitalization and networking, and more than 500 leading intelligent manufacturing demonstration factories have been built. The

production efficiency, product yield and energy resource utilization rate of manufacturing enterprises have been improved significantly, and the maturity level of intelligent manufacturing capability has been improved significantly.

Second, the supply capacity has been significantly enhanced. The technical level and market competitiveness of intelligent manufacturing equipment and industrial software have been improved significantly, with market satisfaction rates exceeding 70% and 50% respectively and cultivated more than 150 intelligent manufacturing system solution suppliers with high professional level and strong service ability.

Third, the basic support is more solid. Building a number of intelligent manufacturing innovation carriers and public service platforms. Building a standard system and network infrastructure to adapt to the development of intelligent manufacturing. Completing the revision of more than 200 national and industry standards and building more than 120 industrial Internet platforms with industry and regional influence.

At the same time, around the four aspects of innovation, application, supply and support, the *Smart Manufacturing Development Plan for the 14 Five-Year Plan* has deployed six special actions, including tackling key problems in intelligent manufacturing technology, building intelligent manufacturing demonstration factories, upgrading industries intelligently, innovating and developing intelligent manufacturing equipment, enhancing software breakthroughs , and piloting intelligent manufacturing standards.

The *Smart Manufacturing Development Plan for the 14th Five-Year Plan* takes the development of advanced intelligent manufacturing industry as the core goal, and lays out the promotion path of manufacturing power. Based on the strong support of the state for intelligent manufacturing, China's intelligent manufacturing industry has maintained a rapid growth rate. At the same time, China's intelligent manufacturing industry has entered a new stage of multi-level promotion and collaborative innovation from the initial concept popularization and pilot application.

The Times Dividend of Intelligent Industry

The word "potential" in Chinese characters is closely related to the word "dividend" in meaning.

The ancient word for "potential" means that a sphere is about to roll down on the slope of a high mound. If the intelligent industry is regarded as a sphere, the dividend is an inclined plane. The longer the inclined plane and the greater the vertical height, the smoother the ball rolls and the more energy it generates.

What is the dividend that constitutes the "inclined plane" of China's intelligent industry? Why is China's development of intelligent industry "following the trend"?

In the previous manufacturing era, China's labor force quantity and labor costs were huge dividends of the times. After entering the era of intelligence, China's labor dividends have not actually disappeared, but there are some changes in the expressions, which is called the talent dividend, that is, the engineer dividend.

There are a large number of dedicated engineers in China. Professor Gan Jie of Cheung Kong Graduate School of Business once said, "Looking for hardware and software engineers related to the IoT in Munich, you may not find anyone after the job advertisement is sent out for half a year. If you are in the United States, you should either pay high salaries to find talents from big companies such as Apple or find old engineers in their fifties or sixties. Siemens has made an internal report that Chinese engineers work twice as many hours a year as foreign engineers."

Gong Ke, president of the World Federation of Engineering Organizations, once said, "In China, there are more than 1.4 million graduates with ordinary bachelor's degree in engineering every year, and there are currently 76 sponsoring member units in China Engineers Federation, which will unite more than 42 million engineering and scientific talents in China. The engineer dividend has become an important force to promote the high-quality development of China's economy, and it also enables China to become an important driving force for the development

of world engineering."[1] China joined the WTO in 2001, and the massive low-cost labor force was the key comparative advantage to promote China to become the world's factory, which is widely called China's "demographic dividend 1.0". Twenty years later, when the labor force in China's factory floor decreased, and some people began to question the subsequent vitality of Chinese manufacturing, industrial manufacturing ushered in the era of intelligent manufacturing that shifted from relying on workers to relying on automated equipment. The children of the workers who used to sweat in the workshop got a good education and became "talent dividend 2.0" in the field of engineering science and technology.

From artificial manufacturing to intelligent manufacturing, from demographic dividend to talent dividend, the self-evolution ability of Chinese manufacturing has amazed the world.

In addition to the talent dividend, we can also look for the times dividend of China's intelligent industry from the above-mentioned keywords of the strategic layout of intelligent industries in the United States and Germany.

"China's manufacturing volume is large. From 2012 to 2020, China's industrial added value increased from 20.9 trillion yuan to 31.3 trillion yuan, of which the added value of manufacturing increased from 16.98 trillion yuan to 26.6 trillion yuan, accounting for 22.5% of the world's total. Up to nearly 30%. China's industry has 41 major categories, 207 middle categories and 666 small categories. It is the country with the most sound industrial system in the world. Among the 500 major industrial products, the output of more than 40% products ranks first in the world. "[2] This is the industrial dividend of China's manufacturing fundamentals, and occupies an important position in the world.

"Digitization" refers to a "digital complex" composed of a series of digital technologies such as big data, AI, mobile Internet and cloud computing. China has the largest population and the largest market in

---

1   Li Kun, Song Rui, *Outlook*, Gong Ke: The Engineer's Dividend Needs Re-recognition", May 31, 2021.

2   Zhou Yi, Sun Mingwei, *The Paper*, "Ministry of Industry and Information Technology: China's Added Value of Manufacturing Industry Ranked First in the World for 11 Consecutive Years", September 13, 2021.

the world. The popularity of the Internet makes China the country with the largest amount of data in the world, so it has considerable advantages in the amount of data required for the development of intelligent industry. This is the most critical data dividend when the era of intelligence comes.

At the same time, as Henning Kagman pointed out, "Digital infrastructure construction is the key to the next industrial revolution. Network construction and bandwidth expansion are very important".[1] By the end of June, 2021, China has built 961,000 5G base stations, and the 5G network has covered all prefecture-level cities, more than 95% of counties and 35% of villages and towns in China. This is the infrastructure dividend of the development of intelligent industry in digitalization.

From demographic dividend to talent dividend, this is the advance of dividend, which makes the demographic dividend in the previous generation of manufacturing industry drive the overall development and status of manufacturing industry and lays the foundation for new intelligent industries. This is to exchange dividends for dividends.

The development of China's intelligent industry can enjoy many dividends, and also suffer many shortcomings. Fortunately, the overall trend is like a ball starting from an "inclined plane", with increasing potential energy and better momentum.

If we broaden our horizons and look at the intelligent industry itself from a higher and wider perspective, the intelligent industry is the dividend of the times. Whoever values the intelligent industry and develops it will enjoy more opportunities and possibilities in its development.

---

1 Hu Xiaobing, *Xinhuanet*, "When the Machine Is Connected with the Network, the Next Industrial Revolution Will Take Place", April 9, 2014.

Building a "Smart Town" and Upgrading Chongqing's Competitive Advantage

Chongqing's industrialization journey has gone through hundreds of years.

In the past century, China has experienced three large-scale relocations of manufacturing industry. During the Anti-Japanese War, coastal industries moved inward. After the founding of PRC, the core industries moved inward during the third-line construction period. Under the background of globalization, China's coastal industrial structure adjustment promotes the inward relocation of electronic industry. Chongqing has seized the opportunity in every relocation.

At the end of 2017, Chongqing proposed to lead the industrial transformation and upgrading with big data intelligence and identified 12 key development areas of intelligent industries such as big data, AI and integrated circuits. Then the 2018 Smart China Expo was held in Chongqing and settled permanently. Since then, Chongqing's manufacturing industry has added intelligent factors and intelligent wings, and started a new take-off.

Chongqing's manufacturing transformation journey has started again, speeding up the construction of a "smart town" and completing the upgrade of Chongqing's competitive advantage, which is destined to be even more magnificent.

Digital Industrialization

At the opening ceremony of Smart China Expo 2021, Chongqing once again sent a clear signal: persist in digital industrialization and dig-

italization of industries, actively promote the deep integration of digital economy and real economy, comprehensively strengthen strategic planning and deployment, speed up the construction of new infrastructure, make every effort to build a smart town and carefully cultivate a smart city.

The definition of digital industrialization is the products and services brought by digital technology, such as electronic information manufacturing industry, information communication industry, software service industry and Internet industry, which emerged after digital technology.

In terms of digital industrialization, efforts are made to build the whole industrial chain of "chips, LCD, smart terminals, core components, and IoT", which carries five huge industrial clusters: integrated circuit industrial cluster, new display industrial cluster, intelligent industrial cluster, core device industrial cluster, industrial Internet and software industry, and other information service industry clusters.

The cluster of "chips, LCD, smart terminals, core components, and IoT" has landed in Chongqing and blossomed everywhere.

Chip: Chongqing is one of the earliest cities in China to develop large-scale integrated circuits, which has initially established a whole process system of "IC design—wafer manufacturing—packaging testing and raw material matching". There are 63 key integrated circuit enterprises and nearly 40 design enterprises. Focusing on enterprises, the 12-inch wafer production line built by China Resources Microelectronics (CR Micro) is leading the enterprise to achieve a breakthrough in manufacturing capacity; the joint microelectronics center has released three sets of process PDKs including 130nm complete silicon photonics process PDK...

LCD: In recent years, Chongqing has initially built a whole industry ecosystem of "glass substrate—liquid crystal panel—display module—complete machine", forming a development pattern of "one core and three parks", and its output value ranks among the top in the country. The main enterprises are BOE, Huike Corporation, Laibao Technology in Chongqing, Corning Incorporated, including two important production lines of BOE's 8.5th generation LCD panel, Huike Jinyu's 8.6th generation LCD panel, BOE's 6th generation AMOLED (flexible) panel

and two key projects of Konka Semiconductor Optoelectronics Industry park.

Smart Terminal: Chongqing's intelligent terminal industry has formed an industrial system of "diverse brands, many OEMs, various supporting facilities, and various types of products". The notebook computer industry has gathered six global OEMs, including Quanta, Inventec, Compal, Pegatron, Wistron, and Foxconn, and has been the world's largest notebook computer production base for seven consecutive years; 3 mobile phone brands ranked among the top 6 in terms of global shipments, accounting for about one-tenth of the world's output; more than 1,000 smart terminal components companies such as BOE, Laibao Hi-Tech Co., Ltd, Lianchuang Electronic Technology Co., Ltd, and Huike Corporation have settled in. There are more than 220 major enterprises. The annual output of PC accounts for about 30% of the world's total, mobile phones account for about 10% of the world's total, and smart watches exceed 20% of the world's total. The new terminals are located in the Xiaomi ecological chain enterprise Fengmi Technology, which produces intelligent projection terminal products, and Fmart vacuum cleaner, which focuses on cleaning robots for household floors and facades.

Core Component: In the field of smart sensors, there are about 60 companies in Chongqing, and the sensor industry scale is nearly 20 billion yuan. The sensor industry ecological chain of "material + design + manufacturing + packaging testing + integration" has been initially formed. The sensor field has achieved full coverage. In 2020, Beibei Park of Western China (Chongqing) Science City was awarded the city's only "Chongqing Sensor Featured Industrial Base", and the *2020 CCID Consulting Top Ten Sensor Park White Paper* was released at the 2020 World Semiconductor Conference Global Sensor and IoT Industry Innovation Summit. In the white paper, Chongqing (Beibei District) sensor characteristic industrial base ranks ninth in the country, and it is also the only selected industrial park in the western region. At the same time, Chongqing is committed to improving the ecology of the new energy and intelligent networked automobile industry, developing core components such as "three big power systems" including power battery, drive motor, electronic control (motor controller), and "three

small electric devices" including DC / DCconvertor, on-board charger, power distribution unit and intelligent control systems. With Liangjiang New District and Yongchuan District as the core areas, Chongqing has formed the "1+10+1000" advantageous auto industry clusters with Chang'an system as the leader, 10 leading vehicle companies as the backbone and thousands of supporting companies as the support.

IoT: The development trend of Chongqing Industrial Internet is doing well . As one of the five top nodes in the country for identification analysis system of industrial Internet, Chongqing has launched and connected 19 secondary nodes, the identification registration volume has exceeded 4.1 billion, the daily identification analysis volume has exceeded 14 million, and the data improvement speed ranks first in the country. Chognqing has built 49,000 5G base stations, formed an efficient and stable network system covering major industrial manufacturing areas, and become the first batch of 5G-scale network pilot cities in the country. Among the 15 cross-industry and cross-domain industrial Internet platforms selected by the Ministry of Industry and Information Technology, 11 platforms are located in Chongqing. Chongqing has gathered nearly 200 enterprises in platform services, solutions, and big data services, has promoted nearly 100,000 enterprises to "go to the cloud and platform", and has accumulatively implemented more than 3,000 intelligent transformation projects... In terms of specific projects, CISDI Group Co., Ltd, which is rooted in Chongqing, launched the first industrial Internet platform in the steel industry. The automobile and motorcycle parts and electronic manufacturing industry Internet public service platforms built by the Chongqing Branch of the China Industrial Internet Research Institute and the Chongqing Branch of the National Industrial Big Data Center have been put into use.

The 14th Five-Year Plan of Science and Technology Innovation in Chongqing puts forward that we should focus on the construction of a "smart town" and a "smart city", focusing on R&D and innovation on the one hand, and complementing the chain on the other, and strive to build a whole industrial chain of "chips, LCD, smart terminals, core components, and IoT". Li Xiaohong, president of the Chinese Academy of Engineering, once said, "About 30 academicians of the Chinese Academy of Engineering have made two rounds of argumentation, giv-

ing high praise to this plan, which is considered as a bellwether or even an engine of technological innovation in Chongqing in the future."

As the five cores of Chongqing's intelligent industry, the rapid development of the whole industrial chain of "chips, LCD, smart terminals, core components, and IoT" in recent years has created a new development pattern and competitive advantage in Chongqing.

Industry Digitalization

As an important manufacturing base, Chongqing has a profound industrial foundation. Among the 41 industrial categories in China, Chongqing has 39, which can be described as complete industrial categories. The digital transformation of Chongqing's manufacturing industry has a good industrial foundation, and "industrial digitalization" also brings new strategic opportunities to Chongqing.

The magic of digital industry has taken effects in Chongqing's automobile and motorcycle manufacturing industry.

"In Chongqing Liangjiang Factory of Chang'an Automobile, an agile and intelligent multi-model flexible production line has been built through the whole process of manufacturing data empowerment, and it takes only 52 seconds for a high-quality car to get off the assembly line. In Liangjiang Smart Factory of JinKang Seres, the production spraying system that supports customers to customize the color of the whole vehicle can be switched in only 15 seconds."[1]

Zongshen Group, a motorcycle manufacturer, launched the Humi Net industrial service platform and took the lead in arranging it on Zongshen motorcycle assembly line 1011.

" In this production line, the robot arm shuttles back and forth, and the sensor detects the products in real time without manual operation. Automatic data collection is realized in all links from online to packaging, and the back end can independently arrange the production plan accordingly. The number of employees in this production line has been

---

1 *People's Daily*, "Intelligence Expo 2021: Digital Economy Helps Chongqing's High-Quality Development", August 21, 2021.

reduced by half, the per capita output has been increased by 2.2 times, and the ability of automatic error correction and error prevention has also been increased by 10.6 times, so that an engine can be off the assembly line in 10 seconds on average." [1]

Such "industrial digitalization" transformation has been carried out in Chongqing in large numbers. According to the data from the Chongqing Economic and Information Commission, in 2021, Chongqing promoted the implementation of 1,295 intelligent transformation projects (4,075 in total, directly driving industrial investment of more than 60 billion yuan) and identified 38 intelligent factories (105 in total) and 215 digital workshops (574 in total). [2]

In addition, it is estimated that the output value of more than 1,500 regulated enterprises in the city that have implemented intelligent transformation has been increased by 46.8% on average, contributing more than 60% to the industrial output value of the city. The production efficiency of demonstration projects in digital workshops and smart factories are increased by 59.8% on average.

The digital economy contributes more and more to Chongqing's regional economy. According to statistics, from 2018 to 2021, the digital economy of Chongqing was increased by 13.7%, 15.9%, 18.3% and 15% respectively, maintaining a high growth rate all the year round. In 2021, the added value of Chongqing's digital economy exceeded 730 billion yuan, accounting for 27.2% of the regional GDP, ranking the first phalanx of the national digital economy.

At the 2021 Smart China Expo, the CICS-CERT released the *Data Map of Chongqing's Integration of Industrialization and Digitalization (2021)*, which showed that Chongqing's integration of industrialization and digitalization ranked first in the central and western regions for many years, and the contribution rates of high-tech industries and strategic emerging industries to industrial growth reached 37.9% and 55.7%

1   Li Jianchang, He Zongyu, *Xinhua Daily Telegraph*, "'Smart Park' Condenses Chongqing's 'Smart Manufacturing', Smart Industry Reaches Trillions in Three Years", December 23, 2019.
2   Liang Haonan, *CQNEWS*, "Focus on Intelligence, Digitization and Greening: Chongqing Will Accelerate the Pace of 'Four Changes' in 2022 to Boost Effective Industrial Investment", April 27, 2022.

respectively.

The report also draws attention to the fact that 14.2% of Chongqing manufacturing enterprises initially have the foundation to explore intelligent manufacturing, and the integration and development of five major industries, such as automobile and motorcycle, electronic information, assembly, materials and chemicals, and medicine, have achieved remarkable results.

This shows that Chongqing's key industries and leading manufacturing enterprises have been integrated into the tide of the intelligent era, upgraded their competitive advantages, and are about to lead Chongqing's intelligent industry to set sail.

Section IV

The Past Five Years, the Key to the Future Decades

If the time went back to August 26 to 28, 2019, in many reports about the 2019 Smart China Expo, we would find that "the forefront" appears very frequently.

This phase is from the important speech delivered by Liu He, Member of the Political Bureau of the CPC Central Committee and Vice Premier of the State Council, at the opening ceremony of the Expo. He pointed out that Chongqing's economic and social development was driven by all positive factors and made remarkable achievements. The development of Chongqing's intelligent industry has been in the forefront in China.[1]

In the history of Chongqing's intelligent industry, the year of 2019 is a link between the past and the future. First, this is the second year that Smart China Expo has permanently settled in Chongqing, and the aggregation effects between the Smart China Expo and Chongqing's intelligent industry were intensifying. Second, as mentioned above, the development of Chongqing's intelligent industry entered the forefront in the country in 2019, when the development conditions became more and more complete, and the development momentum became stronger and stronger.

Since Chongqing held the Smart China Expo in 2018, the development of Chongqing's intelligent industry entered the fast lane in 2019, and then in the five years to 2022, Chongqing has seized the historical

---

1 Han Zheng, Sean, *Chongqing Morning News*, "The Development of Chongqing's Intelligent Industry Is in the 'Forefront': The Next Goal Is to Build a Smart Town and A Smart City with Intelligence", August 30, 2019.

opportunity of building a "smart town" and a "smart city" at a high level, pointing out the direction and laying a solid foundation for the development in the coming decades.

"Wisdom" and "Practice" of the Smart China Expo

We can never underestimate the impact of an expo on a city or even a country, nor underestimate the complementary relationship between an expo and related industries in the city or country.

Hannover Messe has made great contributions to Germany's becoming a world manufacturing power. The annual International Consumer Electronics Show held in Las Vegas not only expands the global influence of the United States in the field of consumer electronics and technology, but also promotes the deep development of the United States in this field with the global knowledge, talents, products and technologies.

2022 will be the fifth year that the Smart China Expo has settled in Chongqing. In five years, it may be too short to sum up the impacts of the Expo on China's smart industry; but in five years, it is enough to show the changes brought by the Expo to the city of Chongqing and the key foundation for future development.

It is divided into two aspects: "wisdom" and "practice".

"Wisdom", according to its own meaning, namely the ideas of experts and scholars, the technical routes and genres of practitioners, and even the presentation of conceptual products and cutting-edge products, has brought "wisdom" to Chongqing and the intelligent industry. This "wisdom" is the direct "wisdom" to promote the development of intelligent industry in Chongqing in the coming decades.

There is also indirect "wisdom", which means "road" and "consensus". What road? Vigorously developing the road of intelligent industry. What consensus? The consensus of vigorously developing intelligent industry.

To some extent, "road" and "consensus" are crucial.

"Road" determines the direction, relates to a city's development strategy, and is the top-level design of urban development. "Consensus"

determines how much strength can be gathered for this road. After five years' publicity and education by the Smart China Expo, Chongqing has been deeply branded with intelligent manufacturing in the hearts of citizens and people all over the world who are concerned about the intelligent industry. This is the consensus, which will attract local and international industry stakeholders to conduct relevant research, investment, and entrepreneurship in Chongqing, thus promoting the rapid development of the industry.

The above mentioned is the "wisdom" part and the following is the "practice" part , both of which have been brought to the future development of Chongqing in the past five years by the Smart China Expo.

In 2018, the official data of the Smart China Expo showed that 501 major projects were signed according to the combination of on-site centralized signing and off-site special signing, with a total investment of about 612 billion yuan. Among them, 56 projects were signed on site with an investment of 190 billion yuan, including 36 contracted projects in Chongqing from 20 districts, counties and development zones with an investment of 107.6 billion yuan. 10 international and neighboring provinces and cities signed projects from Singapore, Hubei and other regions with an investment of 10 billion yuan. 10 strategic cooperation projects signed by Chongqing Municipal Government and relevant departments with well-known enterprises and institutions at home and abroad, with an investment of 72 billion yuan. In addition, Chongqing's districts, counties, development zones, and neighboring provinces and cities signed 445 off-site special projects with an investment of 422 billion yuan, covering all industries of big data intelligent innovation at home and abroad.[1]

It can be seen from the signing data in 2018 that all districts and counties in Chongqing have launched comprehensive cooperation with international, provincial, municipal enterprises and institutions in the intelligent industry. Although these contracted projects may not only be promoted by the Smart China Expo, we can't ignore the platform, si-

---

[1] Qu Hongrui, Liu Hanshu, *Shangyou News*, "About 501 Major Projects Were Signed in the First Smart China Expo, with a Total Investment of about 612 Billion Yuan", August 23, 2018.

phon and inspiration of it.

From 2019 to 2021, according to the statistics of signing projects published in four Smart China Expos, Chongqing signed a total of 1,194 related major projects with a total investment of over 1,952.5 billion yuan.

As we can expect, with the focus on the new generation of information technology, intelligent manufacturing and intelligent services, more than 1,300 projects in key areas of intelligent industries and intelligent applications covering integrated circuits, IoT, big data, new energy and smart cars, intelligent hardware, intelligent equipment, intelligent factories, intelligent headquarters, and intelligent logistics with an investment amount of nearly 2 trillion yuan, gradually took root in Chongqing, releasing their effectiveness, building a huge foundation of intelligent industries and playing an aggregation role.

This is the "wisdom" and "practice" brought by the Smart China Expo to Chongqing. Of course, in the past five years, it was not only the Smart China Expo that marked and led the development of Chongqing's intelligent industry. In fact, the key to future development lies in the top-level design of policies, the planning of industrial chain and the creation of innovative atmosphere.

The Key to Future Competition

Intelligent industrial development is inseparable from scientific and technological innovation.

According to the data of the fourth quarter of 2021, Chongqing has built 21 state-level enterprise technology innovation centers in the field of intelligent industry, and the number of enterprises with R&D institutions is about 1,500. A number of high-end R&D testing carriers, such as The West Institute of China Academy of Information and Communications Technology (CAICT), have successively settled down. A number of high-end innovation platforms such as United Microelectronics Center and Intel FPGA China Innovation Center have also been completed and put into use.

At the same time, Chongqing has successively introduced many

famous universities from home and abroad, such as Peking University, Tsinghua University, Chinese Academy of Sciences, Shanghai Jiaotong University, Harbin Institute of Technology, Huazhong University of Science and Technology, Hunan University, University of Electronic Science and Technology of China, and National University of Singapore. These well-known universities have settled in Chongqing in different forms, such as research institutes, research centers and science and technology innovation centers, which have strengthened Chongqing's scientific and technological innovation strength and made up for the shortage of Chongqing's talent resources.

According to the news in January 2022, the Ministry of Science and Technology supported Chongqing to build a national demonstration zone for the transfer and transformation of scientific and technological achievements. This is the third national demonstration zone for the transfer and transformation of scientific and technological achievements approved by the Ministry of Science and Technology since the 14th Five-Year Plan. Based on the framework of the demonstration zone, Chongqing will optimize the source supply of scientific and technological achievements, improve the level of scientific and technological achievements in the pilot test, and accelerate the iterative upgrading of industries by accelerating the establishment of scientific and technological achievements transformation system and service system, developing the innovation ecosystem around universities, creating a high-level business incubation platform, and creating regional collaborative transformation of scientific and technological achievements.

This is the innovation driving force of Chongqing's intelligent industry in the future.

In addition, the key to Chongqing's smart future lies in digital infrastructure and top-level design based on advantageous industries.

According to the *Science and Technology Daily*, "Chongqing takes 5G as its starting point, actively lays out the innovation and development of the IoT and consolidates the industrial network foundation. Chongqing has become one of the first batch of 5G-scale networking pilot cities in China, with a total of 53,000 5G base stations and a 5G industrial system. Chongqing has gradually improved the infrastructure of the top-level node (Chongqing) of national industrial Internet iden-

tification analysis, with the number of identification registrations and cumulative analysis ranking first in the country. Chongqing has undertaken the national "Spark Chain Network" blockchain super node and established a service system including 10 core big data operators and 20 large-scale cloud platforms, leading the western region in computing power.[1]

The construction of 5G base station, the launch of top-level nodes of industrial Internet identification analysis, and the computing capability of cloud platform directly determine the data sharing capability, transmission capability and application capability of smart factories in a region.

The top-level design based on advantageous industries and emerging industries determines the focus for the future development of intelligent industries. These top-level designs are mainly reflected in *The 14th Five-Year Plan (2021—2025) for High-quality Development of Chongqing's Manufacturing Industry* published in August 2021.

The plan proposes that Chongqing will continuously enhance the international competitiveness of pillar industries such as electronics, automobile and motorcycle, equipment manufacturing, consumer goods and raw materials, and will focus on building a number of strategic emerging industrial clusters with national influence, involving the new generation of information technology, new energy and intelligent networked automobiles, high-end equipment, new materials, biotechnology and green environmental protection.

Among them, in terms of the new generation of information technology, Chongqing will build important power semiconductor devices, flexible ultra-high definition displays, new intelligent terminals, advanced sensors and intelligent instruments and meters, network security industrial base and China's famous software city in order to meet the needs of building "smart town" and "smart city".

As for new energy intelligent networked vehicles, Chongqing will give full play to the comprehensive advantages of fuel vehicle produc-

---

1 Yong Li, *Science & Technology Daily*, "Point-to-Point Combination to Promote the Evolution of Manufacturing Industry: Chongqing Has Evolved into an 'Smart Town'", August 30, 2021.

tion capacity, parts supporting system and integrated circuit ecology, promote the deep integration of new energy vehicles with information and communication, energy and transportation, and build a domestic leading power battery industry base, hydrogen fuel cell application demonstration base and domestic advanced automotive electronics industry base.

In terms of high-end equipment, Chongqing is going to integrate the sensors, communication modules and other components into the whole machine implanted in application of the new materials such as non-ferrous alloys and synthetic materials; and build nationally important mid-to-high-end CNC machine tools, urban rail transit vehicles, new energy equipment industry bases and the leading industrial robots, additive manufacturing equipment industrial base in the west.

In terms of new materials, facing the urgent needs of industrial development and major engineering construction, Chongqing will build an internationally competitive industrial base of polyamide materials and polyurethane materials and an important national industrial base of advanced non-ferrous alloy materials, glass fibers and composite materials.

In terms of biotechnology, facing the needs of major disease discovery and residents' health management, Chongqing will speed up the listing of biopharmaceuticals, promote the upgrading and development of medical devices, chemical raw materials and preparations, and modern Chinese medicines, and build a first-class biomedical industry cluster in China.

The development planning of strategic emerging industrial clusters is a newer and more intelligent evolution based on the original pillar industries, and the original pillar industries are inseparable from intelligent empowerment, which complement each other, thus forming a superior industry with more cluster effects and richer industrial chain value.

At the main forum of the 2021 Smart China Expo, Wan Gang, Vice Chairman of the National Committee of the Chinese People's Political Consultative Conference and chairman of the China Association for Science and Technology, revealed in his speech that in 2020, the scale of China's AI industry has reached 303 billion yuan, a year-on-year increase of 15%, which is higher than the global growth rate. Finally,

he placed high expectation on Chongqing, hoping that Chongqing will become a highland of inland intelligent industry and digital economy development, and a gathering place for high-level international open cooperation.

In the past five years, through the introduction of "wisdom" and "practice" by the Smart China Expo, through the infrastructure construction in the intelligent era, the stimulation of innovation and the transformation of achievements, the "intellectual change" of the original industry and the creation of new industries, as well as clear path planning and strong implementation, Chongqing, as a highland of intelligent industry and digital economy development, and as a gathering place of high-level international openness and cooperation, is taking shape.

Part III

Intelligent Life: Intelligent Core for Life

The logic of digitalization is integrated into the society, and the smart algorithms cover the city. Once the kernel of intelligent life is started, digital civilization is also integrated into people's beautiful imagination of life, so the whole city's life becomes more colorful with the "smart office", "smart shopping", "smart home" and "smart medical care".

No matter how the city evolves, no matter how the technology changes, the protagonist of the smart city is always the citizens who live in it.

Section I

Intelligence and Digitalization: Permeate Every Corner of Human Life

Occasionally, someone on the Internet puts forward a hypothesis: What would people's lives be like without AI?

Without AI, urban traffic would be paralyzed, online car-hailing would not be available, and the operation of public transport would be blocked. Without AI, commercial materials would not operate smoothly, the market supply and demand would be sharply unbalanced, and the purchase of living materials would no longer be convenient. Without AI, mobile payments would not be available, the cash age would return, and many new lifestyles relying on mobile payments would disappear.

Artificial intelligence has been closely linked with all aspects of human life. Whether it is economic development, urban governance, or people's livelihood, it is inseparable from AI.

It's not difficult to feel the development of science and technology and the arrival of digitalization. The constantly updated intelligent products are flooding into our lives, and the convenient experience brought by these intelligent technologies to our lives is increasing year by year at the rate visible to general public. All aspects of public life including "smart office", "smart shopping", "smart home" and "smart medical care" have become better because of the underlying support of AI.

The continuous evolution of intelligence and digitalization has become an important force affecting the development of human society and has penetrated into every corner of human life.

"Smart Mobility" Creates a New Pattern in the Intelligent Era

From walking to horseback riding, ancient human beings depended on external force to avoid travel fatigue. From steam power to fuel power, modern industry increased travel speed with energy. With the rapid development of science and technology, aircraft and high-speed railway can make people travel thousands of miles a day. The progress of science and technology has greatly reduced the physical labor of human beings and greatly increased the travel speed.

However, in modern cities with tens of millions of populations, the demand for ultra-high-density mobility and the high-speed pace of life are putting forward higher requirements for human mobility: reducing time consumption and mutual interference.

In order to meet these new demands, AI provides the best mobility solution.

When we go to work in the morning, we don't have to rush to catch the bus. We can browse the bus arrival information through the app while having breakfast, and go out calmly and set foot on the required bus on time. When driving, we don't have to worry about road congestion, since intelligent navigation updates the road condition information in real time and optimizes the route. These intelligent experiences of mobility have become people's daily life.

As the main driving force of urban development, traffic has an important influence on the flow of production factors and the development of urban system and is one of the decisive factors of urban prosperity and decline. In the era of rapid development of AI, people have enjoyed the convenience brought by smart technology in all kinds of mobility and life scenarios.

At the Huawei booth of the 2021 Smart China Expo, the theme was "Bring digital world to every person, home and organization for a fully-connected, intelligent world". In terms of smart new mobility, large-scale traffic scenario solutions such as Huawei Smart Airport and Smart Urban Rail were unveiled at the Expo booth in turn. With digital and intelligent solutions such as "one-map of operation control", "face-ID for travel", "airport smart brain" and urban-rail cloud platform, visitors

will be outlined with new scenarios of smart travel in the future.

In practical application, the "Smart Airport" ecosystem that Chongqing Jiangbei Airport is trying to build is known as the "model support" for Chongqing's smart city construction. Here, from check-in, security check to boarding, passengers can complete the whole process of "paperless" flight. Passengers can achieve international self-service customs clearance in 9 seconds[1] and walk around the airport with face recognition. In addition, as one of the first six pilot airports in China, Chongqing Jiangbei Airport has also launched the baggage tracking and query of the "PEK T3-CKG" route operated by Air China and started the baggage data sharing service.

At the 2021 Smart China Expo, Chang'an Automobile also showed the "unique technologies" of smart mobility. Zhu Huarong, chairman of Chang'an Automobile, described the strategic layout of "New Car and New Ecology" to all circles at the summit forum of the Smart China Expo. At its booth, Chang'an Automobile released a number of "black technologies" for smart mobility, such as APA7.0 remote unmanned valet service system, which can launch the unmanned autonomous queuing parking function outside the garage for car owners. The world's first electric drive high-frequency pulse heating technology can solve the range anxiety in winter and can also be heated to 20°C in 5 minutes in an extremely cold environment of -30°C, improving the low-temperature running performance of trams, and with the waste heat recovery technology to increase the cruising range by 40—70 kilometers.

The "New Residence" Era Adds Color to Life

The development of smart home in China has gone through several stages: "from budding, creation, wandering, evolution to outbreak".

In the budding period of 1990s, smart home was still in a stage of concept learning and product cognition, and there were no professional smart home manufacturers at that time. From 2000, the first-tier cities

---

1 Li Jing, *Shangyou News*, "Building a 'Smart Airport', Chongqing Jiangbei Airport Will Do So", April 15, 2021.

such as Shenzhen and Shanghai began to have perfect smart home marketing and technical training systems. During the evolution period from 2011, the volume growth of smart home industry has entered a golden development period. After 2020, the smart home industry began to enter an explosive period.

At present, with the gradual landing of smart single products, the capacity and penetration rate of China's smart home market are in an elevation period. The upstream and downstream industries such as R&D, design, production and sales of smart homes have attracted many enterprises to enter the market, supporting a complete industrial chain, and enabling the smart home industry to develop rapidly on a large scale. The development of AI, IoT, cloud platform and other technologies began to pave the way for the smart home industry. Driven by technology research and development and consumer demand, smart home has evolved into the "new residence" requirement of whole-house intelligence.

Different from the traditional residence, "new residence" focuses on the whole-scenario experience, that is, the whole-house intelligence. According to all living scenarios, it provides users with solutions covering all needs, as well as partial renewal and renovation of kitchens, bedrooms, bathrooms and other spaces.

Chinese families' interest in smart homes has supported a huge market. According to the *2022—2027 Smart Home Industry Deep Survey and Future Development Trend Forecast Report* released by ChinaIRN, China will become the largest consumer of smart home market in the world, accounting for 50%—60% of the global smart home market consumption share. AskCI Resarch predicts that the smart home market in China will reach 651.56 billion yuan in 2022.[1]

According to the data of TianYanCha Professional Edition, there were nearly 160,000 smart home-related enterprises in China in 2021. Besides wholesale and retail enterprises, there were more than 25,000 enterprises engaged in information transmission, software and information technology services for smart homes, and nearly 20,000 enterprises

---

1 ChinaIRN，*2022—2027 Smart Home Industry Deep Survey and Future Development Trend Forecast Report.*

providing scientific research and technical services for smart homes.[1]

In fact, people have taken a closer look at the "home": paying more attention to the healthy property of the house, paying more attention to the operation and management of the community, and demanding for more functional space. When it comes to household products, the function should not only be more comfortable, but also healthier and smarter.

At the 2021 Smart China Expo, all manufacturers came up with the unique technologies of smart home in strategies, technology schemes, and applications. They all stroved to bring consumers into the era of "new residence" based on the industry's largest smart home scenario ecology.

In the past, smart homes were widely focused on IoT applications in indoor scenarios. At the 2021 Smart China Expo, smart homes presented a more systematic solution for intelligent scenario applications. Smart scenarios have also expanded from home applications to community applications, and from smart homes to smart communities.

In terms of smart home, OPPO was also based on UWB spatial perception technology, which brought a new experience of precise pointing and control of smart home, making the control between smart devices more precise. In China Mobile's "Smart Home" exhibition area, by creating an innovative product system of "one all-optical network with N standard products + multiple application scenarios", we could customize the whole house scenario mode with one click, feeling that home life was "smarter, safer, greener and more fun".

In terms of smart communities, 5 smart communities and 5 property companies, including Jinke Bocui Future, Midea Wanlufu, Longfor Smart Service, and Chongqing Vanke Property, which have brought a convenient and efficient living experience to all aspects of the residents' daily life with various intelligent scenarios, effectively improved the efficiency of property management services, and continually enhanced the residents' sense of gain, happiness and security, were rated as "2021 Chongqing Big Data Intelligent Application Ten Selected Cases of the Big 'Smart Community'".

---

1 *Jiemian News*, "TianYanCha: There Are nearly 160,000 Smart Home Related Enterprises in China, and More Than 30,000 New Enterprises Will Be Added in 2020".

"Digitization" Builds a New Picture of a Better Life

Human science and technology and civilization are based on the wish for good life, and the good wisdom life extended by the good life is being written in a world where everything is interconnected and perceivable. The world we live in is rapidly and deeply stepping into the digital age.

Chongqing is responsible for the dual mission of building the National Digital Economy Innovation Development Experimental Zone and the National New Generation Artificial Intelligence Innovation Development Experimental Zone. In recent years, focusing on the key issues that restrict the innovation and development of digital economy and the new generation of AI, Chongqing has made great efforts to build a new type of infrastructure, security system and policy support system, vigorously carried out scientific and technological innovation, reform pilot and application demonstration, actively promoted industrial cultivation and economic transformation, and is growing into a highland for the development of inland intelligent industry and digital economy and a gathering place for high-level international open cooperation.

The Smart China Expo, which has been held for 4 consecutive sessions, has become a normal business card for Chongqing to show the development of digital economy. Besides the typical characteristics of contemporary cities, Chongqing is also an important window to understand the development of modern digital cities in China.

At the 2021 Smart China Expo, Wan Gang, Vice Chairman of the National Committee of the Chinese People's Political Consultative Conference and chairman of the China Association for Science and Technology, said that the industrial application of AI is expanding day by day, and its application is moving out of the laboratory into the production and living scenarios such as manufacturing, transportation, commerce, law, sales, services and urban management, which has further played an enabling role and effectively promoted the development trend of intelligence, digitalization and networking in various fields.

In the past two years, the COVID-19 epidemic has forced many activities to press the "pause button", but digital information technology

has begun to spread its wings. Online classes, online office work, and online shopping have brought us into a more intelligent era.

Intelligent service is the direct change that digital economy brings to life, and digital life is subverting the traditional way of life. All things can be "cloud", such as cloud new year's greetings, cloud learning, cloud office and cloud travel, and "cloud" has become the standard of digital life.

In the blink of an eye, digital transformation has become the urgent need of all trades and industries, which is not only considered to bring about the improvement of ecological value and economic value, but also a key layout to promote the smarter, more efficient, and higher-quality development of the whole human society.

Chongqing's new smart city operation and management center has connected to more than 200 systems, focusing on creating nearly 30 comprehensive application scenarios. For example, "Yukuaiban"—a mobile government service platform officially launched in Chongqing on November 21, 2018; "Yu Kuaizheng"—Chongqing's integrated, intelligent and digital government work platform; "Yu Kuai Rong"—the main platform for Chongqing's corporate financing big data service; "Yukang Code"—providing health information declaration, dynamic health status display, and migrant work health certificate declaration for residents in Chongqing to facilitate travel. These intelligent and innovative applications are bringing great convenience to the urban life, production and operation of Chongqing citizens, and making urban management more intelligent. Smart access control, smart kitchen, smart water saving, smart temperature control, light adjustment and other integrated smart homes are adding luster to home life. Functions such as swiping face ID / QR code to enter the store, visual search for product information, and checkout in the induction area are applied in unmanned shopping malls, making it more convenient for shopping. The application of technologies such as online payment, face recognition, and intelligent guide service in scenic spots provides more convenient, more powerful and more user-friendly intelligent services for traveling. The intelligent application of "food, clothing, housing, travel, shopping and entertainment" continuously meets the needs of the people for

high-quality life.[1]

It is an ideal picture for science and technology to coexist with human beings, and it is also Chongqing's vision to let digital and intelligent technology enter human life and make human life better.

---

1   *Chongqing Daily*, "Building a Famous Smart City with 'Four Beams and Eight Pillars', Chongqing's Smart City Construction Level Is at the Forefront of the Country", August 18, 2021.

Section II

Digital New Life: Chinese Lifestyle Is Leading the Global Trend

Digital, intelligent, and sustainable development of the city has become an extremely important issue in Community of Shared Future for Mankind.

In recent years, China has made great efforts to build "Digital China" with new national competitive strengths of digital economy, focusing on digital technology, industrial digitalization, rural revitalization, carbon peak and carbon neutrality, medical treatment and health, digital life, and services.

China's digital life is leading the global trend and further marching to a wider world.

Chongqing: Creating a New Way of "Digital Life"

Since 2018, the annual Smart China Expo has become an indispensable weather vane for the outside world to pay attention to the latest progress and development trend of smart cities.

Through the Smart China Expo, we may think about some questions: After more than ten years of development, has the smart city industry entered a new stage? How to make intelligent digital technology nourish urban development like river nourishing life?

In fact, after more than ten years' efforts, Chongqing has continuously accelerated the integration of digitalization and intelligence of traditional industries with the real economy and has become a demonstration area of digital economy in China.

On December 8, 2021, the Chongqing Municipal Government is-

sued *The 14th Five-Year Plan of Chongqing Digital Economy (2021—2025)*, stating that by 2022, the city will gather 100 leading enterprises in digital economy, 500 high-growth innovative enterprises in frontier fields, 5,000 small-and-medium-sized enterprises and innovative teams, and create 10 national-level digital economies. By 2025, the city's total digital economy will exceed 1 trillion yuan.

Specifically, Chongqing aims at the key development areas of intelligent industries such as big data, AI, integrated circuits, intelligent networked automobiles, strives to build a "smart town", effectively promotes the industrial-level leap and high-quality economic development, improves urban governance with big data intelligence, and focuses on building a "smart city". Government services, social governance, and people's lives have become smarter, greener, and more efficient.

In Chongqing, the "smart transportation" based on the Internet and big data has greatly eased the urban congestion. "Intelligent flood control" has greatly improved the efficiency of urban inland river regulation and storage and emergency disposal of drainage and flood control. Furthermore, digital lifestyle has entered ordinary people's families, solving many people's livelihood "pain points", "congestion points" and "difficulties".

Chongqing government convenience service platform "Yukuaiban" ushered in version 3.0 in 2021. In terms of convenience service effectiveness, it realizes "one-click search" and "one-click navigation" of 42 government service halls in the city on app. At the same time, it integrates the workers' harbor set up by 277 business outlets of Chongqing Branch of China Construction Bank to provide convenient services such as drinking water, charging and wireless Internet access for outdoor workers.

In addition, "Yukuaiban" 3.0 provides different service applications for different groups. For example, a special service area is set up for the elderly. It has more than 20 senior-friendly items in five categories including social security, medical care, provident fund, household administration and certification services. The newly-added operating tips, voice assistance and other functions, high contrast and more concise interface style with large font size and large icons bring diversified warm

senior-friendly services.[1]

At the moment when the COVID-19 epidemic has not completely dissipated, "Yukang Code"—Chongqing Health Code has been escorting more than 30 million Chongqing people. The digital power behind it is the safe and reliable technical support of Inspur Cloud. There is a huge data support behind the small QR code, including everyone's health data, medical information provided by the health and disease control department, etc.

Chongqing is well aware that it will no longer be the land and demographic dividend that determines urban development, but the data dividend generated by big data operations and services, which will form various smart applications based on data resources and promote the construction of smart cities. Moreover, Chongqing's digital potential is far more powerful than people think.

China's Digital Life Leads the World

The significant achievement of "Digital Chongqing" construction is the epitome of the rapid advancement of "Digital China" construction. With the "two-wheel drive" of digital industrialization and industrial digitalization, and the mutual integration of real economy and digital economy, China's digital economy is in full swing.

According to the *Digital China Development Report* (*2020*) released by the Cyberspace Administration of China, during the 13th Five-Year Plan period, the development vitality of China's digital economy has been continuously enhanced, and the total number of China's digital economy has leapt to the second place in the world, with the added value of core industries of digital economy accounting for 7.8% of GDP.[2]

The rapid popularization of "digitalization" in China is closely related to Chinese people's open attitude towards digital life. Take online

---

1   Liuqiao, Chenshiyi, *Xinhuanet Chongqing*, "Chongqing Online Government Affairs Service Platform 'Yukuaiban' 3.0 Officially Launched", July 28, 2021.
2   Zhang Yang, *Huanqiu*, "'*Digital China Development Report (2020)*' Released That China's Digital Economy Ranks Second in the World", April 26, 2021.

payment as an example. As early as 2017, China has become the largest digital payment market in the world, and QR code has become a widely used payment method.

On March 5, 2018, the People's Bank of China announced *Report of the Overall Situation of Payment System Operation in 2017*. The data showed that in 2017, mobile payment business maintained a rapid growth, with 48.578 billion online payment businesses, amounting to 2,075.09 trillion yuan, and 37.552 billion mobile payment businesses, amounting to 202.93 trillion yuan.

In just four years, on April 2, 2022, in the *Report of the Overall Situation of Payment System Operation in 2021* issued by the People's Bank of China, there were 102.278 billion online payment services with an amount of 2,353.96 trillion yuan and 151.228 billion mobile payment services with an amount of 526.98 trillion yuan in 2021.

With the deepening digitalization of daily life, the number of online payment services of Chinese consumers has doubled compared with four years ago, and increased steadily. The number of mobile payments has increased to four times in 2017, and the number of mobile payments has reached 2.6 times that of four years ago.

It is worth noting that even if global consumption is affected by the epidemic, it is difficult to shake China's position in the development of digital economy in the world. On August 2, 2021, the *White Paper on Global Digital Economy: A New Dawn of Recovery under the Impact of Epidemic Situation* released by China Academy of Information and Communications Technology (CAICT) showed that the scale of China's digital economy in 2020 was 5.4 trillion USD, ranking second in the world.[1]

Under the epidemic situation, "stay-at-home" has become the daily life of most people. Where the demand is, the supply will go. A large number of people can't go out of their homes and walk beyond the community, and the huge demand for living and consumption is superimposed on the grim situation of epidemic prevention and control. It is

---

1 China Academy of Information and Communications Technology (CAICT), *The White Paper on Global Digital Economy: A New Dawn of Recovery under the Impact of Epidemic Situation*, August, 2021.

necessary to reduce the risk of cross-infection and take into account the practical needs of resuming work and production, so "contactless business" has emerged.

Take the catering industry as an example. China Hospitality Association has jointly launched the first batch of "contactless restaurants" with a number of catering brands. Some large shopping malls and supermarkets have focused on developing online sales, and major express delivery companies have introduced their own contactless delivery standards. Through online transactions, offline fixed-point delivery, and self-service by consumers, the direct contact between buyers and sellers can be avoided. This "contactless commerce" mode is favored by consumers.[1]

"Contactless commerce" not only solves the daily needs of the people in the epidemic, but also promotes the digital economy itself to sprout all kinds of brand-new business models. All kinds of new businesses and services emerge in an endless stream, such as unmanned shelves, cashierless store, intelligent express cabinets, unmanned delivery vehicles, etc., which have greatly changed the retail format and created new employment opportunities.

In addition, new cultural and travel products such as "Cloud Travel", "Online Reading Club" and "Cinema Selling and Takeaway" came into being. In fact, our life has long been inseparable from digital technology. Not only have we been "involved", but we are always "online". "Isolating the epidemic without isolating the economy", it must be admitted that China's digital life innovation is an important force leading the economic and social development and influencing the international competition pattern.

On February 5, 2022, during his official visit to China and attending the opening ceremony of the Beijing Winter Olympics, the then Pakistani Prime Minister Imran Khan said, "Take the fight against the COVID-19 epidemic as an example. Around the whole world, no country can

---

1 Zhengzhihui, *Xinkuaibao*, "Contactless Restaurant! China Hospitality Association, Meituan and Several Catering Enterprises Jointly Promote the Whole Chain 'Contactless'", February 21, 2020.

take such powerful measures as China."[1]

China's Digitalization in the Eyes of the World

Besides driving China's economic transformation and development, "digitalization" also plays an important role in improving China's social management efficiency and promoting the social development process. After several years of great-leap-forward development, "Digital China" is not only an economic concept but permeates every link of social development.

Under the digital wave, all kinds of new paradigms, new services and new models are constantly emerging, and self-employment, sideline innovation and flexible employment are flourishing. In rural areas, a mobile phone and a selfie stick are becoming "new farm tools" for more and more farmers, directly linking agricultural products with the consumer market. In cities, online car-hailing and delivery services are running around the streets. The exploration and innovation of digital China is going deep into every corner of the national economy and people's livelihood, so that the vitality of innovation and entrepreneurship will be generated, and the dividends of convenience services will grow.

"Digitization" has become an important way to improve people's living standards, optimize economic structure and promote social development. China has also become one of the countries with the fastest digital development in the world.

Many overseas media have fully affirmed China's achievements in digital technology development and digital economy. "China has entered a new era of digital industry. In the next few decades, China will become the leader of global digital development." When it comes to China's digitalization, *Fortune* once reported.

On April 13th, 2022, Jonathan Woetzel, McKinsey Global Institute Director and Senior Partner, pointed out that China has a unique and

---

1 Xingxiaojing, Liucaiyu, *Global Times*, "The Then Pakistani Prime Minister Imran Khan Accepts an Exclusive Interview with the *Global Times*: A Confident Country Can Host This Grand Event", February 7, 2022.

mature digital ecosystem, a deep digital consumer base and a rapid digital transformation and development speed. Through a series of data analysis and chart demonstration, Dr. Jonathan Woetzel revealed China' leading position in the global digital revolution, and the pioneers of China's digital revolution stood out from high-frequency services and expanded to all-round acquisition services.[1]

The tide of digital economy is in full swing in China. As a new economic development model, China's "digitalization" has been widely recognized by the international community and has even become a familiar "Chinese business card".

1　Ding Shutong, Xie Zhiyu, Official Website of Peking University, "McKinsey Executives Jonathan Woetzel and Jeongmin Seong Talk About China's Digital Transformation and Its Impact in Asia", April 13, 2022.

Section III

Building a "Smart City", Chongqing Becomes a Smart, Job-Pleasing and Livable City

The city carries the aspiration of human beings for a better life. Every time it evolves and upgrades, it is a manifestation of the progress of human society.

The growth of a city is closely related to the progress of science and technology. With the acceleration of new infrastructure, the accelerated construction of "smart cities" around technology, policy and ecology is becoming a common proposition for every city experiencing the transformation of the technological revolution.

The construction of new smart cities is in full swing. Over the years, Chongqing has been determined to promote the planning and development of intelligent industries. This city of mountains and rivers not only integrates the Internet, IoT, cloud computing and other modern technologies into the concept of "smart city", but also brings the "city of dreams" that can sense, think, evolve and have a temperature into our real life.

The Road to "Smart City"

2021 is known as a key year for Chongqing to accelerate the development of digital economy and speed up the construction of "smart city".

In June 15, 2021, *The 14th Five-Year Plan for the Development of Infrastructure in Chongqing (2021—2025)* proposed that by 2025, a modern information infrastructure system that is efficient and practical, intelligent and green, safe and reliable will be built to provide strong

support for the city's economic and social development to higher quality, accelerate the construction of a modern economic system, help the construction of "smart manufacturing town" and "smart city".

The debut of these two business cards can be traced back to the stage of the 2019 Smart China Expo, when Chongqing resoundingly proposed to focus its efforts on building a "smart manufacturing town" and a "smart city" to harness the momentum in the digital industrial transformation and grasp the initiative of the new round of technological revolution.[1]

This flagging concentration of power takes into account the profound accumulation of the old industrial base in Chongqing, but also determines the future direction of the development of the digital economy in the age of intelligence and maps out a clear path for the long-term development of the city.

As long as we tour around the exhibition hall, it is not difficult to find that the traditional manufacturing industry in Chongqing is being revitalized, and the smart manufacturing deeply linked with the industrial Internet platform is ubiquitous.

As of August 2021, Chongqing has built 200 smart agriculture demonstration bases, promoted the demonstration construction of 350 smart parks, 99 demonstration smart tourism scenic spots, built 191 smart communities, 545 smart property communities, and built 44 smart hospitals, built the city's health and medical big data resource lake and platform foundation. In addition, Chongqing has also focused on creating application scenarios for grassroots governance such as emergency management, smart fire protection, and legal services for villages and communities, as well as application scenarios for people's livelihoods such as smart parking, bright kitchens and stoves, and smart transportation. Smart life is entering all aspects of citizens' lives, bringing enormous and beautiful changes.[2]

Beibei District has built and put into use the first smart minor pro-

---

1 Han Zheng, Chen Xiang, *Chongqing Morning Post*, "Chongqing's Smart Industry Development Has Been in the 'Forefront' and the Next Goal Is to Build a Smart Town and a Smart City", August 30, 2019.
2 *Chongqing Daily*, "Make Life More Exciting! The Construction of 'Smart City' Is Getting Better", August 20, 2021.

tection system in Chongqing, which divides children in distress into three categories, and equips some children with smart watches to prevent major safety risks such as lack of guardianship. Guardians, social workers, and the community can use voice call, health data monitoring, tracking and positioning and other functions, timely grasp the status of children, and provide precise positioning for emergency assistance, reducing the danger that minors may face.

The first smart supervision farmers market in Chongqing—Peninsula Yijing Farmers Market, nearly 100 booths are neatly divided into fresh, live, raw, cooked, dry, wet and other areas, and the large screen of each area scrolls to announce the results of the day's fruit and vegetable sampling. An electronic screen stands in front of each vendor, and the personal information of the operator, the star rating, the price of the food of the day and other information are clear.[1]

The self-driving bus shuttle service on a 5.4-km circular route in Chongqing High-Tech Zone has been officially operated at the end of August 2021. The project has built intelligent facilities and equipment such as smart bus stops, smart crosswalks, smart ramps and smart intersections through C-V2X, 5G, edge computing and other technologies, with vehicle-road cooperation as the core concept. Driverless vehicles can not only smoothly accelerate and decelerate, change direction and make other operations, avoid collision with obstacles and vehicles, but also "get along" with other social vehicles. In addition, the in-vehicle screen can display information about objects around the vehicle and the route the vehicle is expected to travel, which is full of technology. It is understood that the project operation mode is also applied to the High-Tech Zone Xiyong, University City and other areas, effectively solving the people's mobility problems and easing the pressure of traffic congestion.

In front of Changjiahui Shopping Park in Nan'an District, Chongqing, the "Smart Zebra Crossing" project, upgraded by "IoT + Intelligent Sensing" technology, was officially "inaugurated" in October 2021. The project integrates pedestrian flow monitoring and flashing lights, and reminds oncoming vehicles and pedestrians in the two way to pay

---

1 Luo Bin, *Chongqing Daily*, "Smarter Cities, More Comfortable Citizens, Smarter Empowering Chongqing Citizens to Live a Happy Life", July 2, 2022.

attention to traffic safety through the safety island transformation, additional ground light-emitting road studs, light-emitting warning tape and pedestrian flow monitor.

A new scenario of Chongqing intelligent logistics application has been landed in Chongqing Lianglu Cuntan Bonded Port Area. Chongqing FEILIKS three-dimensional warehouse has realized the intelligence and unmanned whole system of image identification, inbound, outbound, and AGV handling vehicles. Automated warehouse storage capacity and inbound speed are 3 times faster than traditional warehouses, outbound speed is also 2.5 times faster than traditional warehouses, inbound and outbound speed up to 960 boxes per hour, and the accuracy rate of reaches 100%.[1]

Illuminating the future of urban development with the light of wisdom. In addition to the main urban areas throughout Chongqing, the districts and counties in the region are also accelerating the pace of building a smart city. Chongqing Yongchuan District introduced at the Smart China Expo that Yongchuan has achieved the debut of the L4G self-driving mid-bus jointly developed by the local King Long Bus and Baidu for the development of smart transportation. In order to create a smart city management, Yongchuan accessed the video monitoring system and established a five-in-one smart management system of perception, analysis, service, command and monitoring. In addition, Yongchuan District also jointly developed a digital twin city decision-making platform with China Aerospace Science and Industry Corporation and Dava Technology to provide decision aids for urban construction and management, spatial planning and fire rescue.

Let "the Near Ones Please, the Far Ones Come", Chongqing's "Job-Pleasing Gene"

More than 2000 years ago, Aristotle once said, "People come to the city in order to live, and stay in the city in order to live better." Today,

1 Gu Li, Zhang Yunqiu, Ran Lili, Tan Li, *Chongqing Daily*, "Wisdom Gathered: Chongqing FEILIKS, Speeding Up Intelligent Logistics", November 18, 2020.

big data intelligence for the improvement of urban quality has injected more "wisdom factors", so that intelligence accurately meets people's needs for a better life.

Suitable-to-work is the most long-lasting and vigorous vitality of a city. A city that can retain talents and make them willingly take root must have the "job-pleasing" genes, such as a long history, good economic environment, convenient transportation, rich cultural heritage, and favorable talent policies.

The youngest municipality, the new first-tier Internet-famous city, the 8D magic city... Chongqing, the city of mountains and rivers with multiple labels, the city of more than two million migrant population, how to make them feel willing to settle down and ensure stable employment? How to better serve the economic and social development? How to improve the livelihood of the people so that "the near ones please, the far ones come"?

Chongqing is surrounded by rivers and mountains, which is not conducive to the development of modern transportation and modern architectural planning. However, after the establishment of the municipality, the economy has grown strongly and the population has continued to gather, growing into a cosmopolitan city at a very fast pace. The city of Chongqing has created too many miracles.

In terms of transportation, Chongqing pressed the "fast-forward" button in 2021, and ran with "acceleration" to achieve "high efficiency". The first 10,000-ton terminal in the upper reaches of the Yangtze River, Chongqing Xinsheng Port opened for operation; strategic, iconic and exemplary major projects in the twin-city economic circle of Chengdu and Chongqing—the Chengdu-Chongqing Central High Speed Railway with a design speed of 350 kilometers per hour started construction; Hechuan to Changshou Expressway will soon be opened to traffic, and the main metropolitan area of Chongqing will enter the "the third-ring era"; Kaizhou Tanjia to Zhaojia section of Chengkou to Kaizhou high-speed rail is officially opened to traffic, and Chongqing "county to county expressway" is just around the corner.

In terms of economy, according to the *2021 Chongqing National Economic and Social Development Statistical Bulletin* released by Chongqing Municipal Bureau of Statistics and Chongqing Survey

Headquarters of the National Bureau of Statistics on March 18, 2022, in 2021, Chongqing ranked 3rd among 11 provinces and cities in the Yangtze River Economic Belt with a two-year average growth rate of 6.1% in gross regional product, 2nd in total retail sales of consumer goods, and 1st in per capita disposable income of urban and rural residents, with Chongqing's supporting role in the Yangtze River Economic Belt increasing. In 2021, the main city metropolitan area, the Three Gorges Reservoir Area in northeast Chongqing, and the Wuling Mountain Area in southeast Chongqing achieved a regional GDP of 2,145,564 million yuan, 489,515 million yuan and 154,319 million yuan, respectively, with growth rates of 8.0%, 9.1% and 7.6%, respectively, over the previous year.[1]

In terms of talent policy, Chongqing took the initiative of "inviting in" and "going out". in 2021. Among them, "Millions of Talents Boost Chongqing" Beijing trip and Sichuan trip and other activities are one of the latest projects of Chongqing talent introduction work of "going out". And for the talent "inviting in" management and other work, Chongqing human resources and social security departments adhere to the classification to promote talent evaluation, establish classification evaluation title assessment "green channel", scientific research project funding "overall rationing system", open competition mechanism to select the best experts and so on, powerfully motivating talents to stand out. It is known that Chongqing talent services were upgraded to version 3.0 in 2021, which can provide 25 kinds of services for Chongqing talent card holders through telephone, website, WeChat and platform.

......

On May 11, 2022, under the guidance of a clearer urban strategy, Chongqing focused on key industries and key areas, and launched an action to accurately introduce urgently-needed engineering talents to the world. Specifically, around the 7 pillar industries and 33 industrial chains in Chongqing, as well as national-level small-and-medium-sized industrial enterprises with the characteristics of "specialization, refinement and novelty", through the combination of "online + offline",

---

1 Zhao Yingzhu, *CQNEWS*, "Total Economic Volume Steps Up to 2.7 Trillion Yuan! 2021 Chongqing Economic Report Card Hides These Highlights", March 18, 2022.

Chongqing has organized targeted talent introduction activities to introduce overseas engineers and outstanding engineers.

In the world's eager eyes, Chongqing, which carries the lead in the upper reaches of the Yangtze River economy, is standing at a new starting point and is going to start a long journey to rewrite history: a scene of the rise of a modern metropolis opens here, and an avenue to the era of intelligent industries is built here.

Create "Chongqing-Style Happy Life"

Locals and out-of-towners alike have their own good reasons to love a city. For the general public, the reasons for being willing to stay in a city lie in economic vitality, and the ability to provide sufficient jobs and other material security. In addition, the pursuit of a good spiritual life has also become an important consideration.

How to determine whether a city is livable or not? A beautiful environment, social security, civilization and progress, comfortable living, economic harmony and high reputation constitute the basic characteristics of a livable city.

Chongqing is a city of mountains and rivers, and the residents of the city are very lucky to live in the scenery. Every visitor could enjoy the soul of the canyon, the night rain in Bashan, and the fairyland of the sea of clouds; the bright lights of Hongya Cave, the hustles and bustles on the lanes and allies; the noodles in the morning, and the hot pot at night.

In Chongqing, we can feel the collision of history and modernity, tradition and fashion, as well as the perseverance and briskness of Chongqing people engraved in their bones. In terms of climate, urban construction, cuisine, scenery and environment, Chongqing is a livable city. This amazing land is also known as the city of history, culture, tourism and civilization by foreign visitors.

The open policy, inclusive disposition, poetry and dreams are in sight, which attracts more and more young people to work and live here, and follow their dreams and realize dreams here.

People with dreams are happy. Happy people have their own ex-

periences, and a life filled with happiness is always desirable. When a city faces the future, it has a common, clear and grand dream, and when people are in it, the sense of happiness has a broader bearing space.

Urban construction is for the people who live in it. Only by allowing citizens to enjoy the convenience of digital life and share the dividends of the intelligent era can the city have a better future. The continuous hosting of the five-year Smart China Expo has not only injected strong innovation and competitiveness into the smart industry in Chongqing, but also brought a full sense of achievement and happiness to the residents of Chongqing.

In recent years, on the one hand, Chongqing has deployed industrial digitization and digital industrialization, promoted the transformation and upgrading of traditional industries, cultivated strategic emerging industries, and achieved high-quality economic development; on the other hand, Chongqing focuses on people's livelihood projects such as living environment, education, and medical care, and uses the achievements of intelligent development to feed back urban construction and improve people's well-being. The "two-highs" —high-quality development and high-quality life, go hand-in-hand, creating a "highway" leading to a happy life for Chongqing people.

The Smart China Expo is in such a way to announce that Chongqing is aiming to build a modern city that is livable, resilient and smart, and striving to make people's life in the city more convenient, comfortable and better.

In fact, Chongqing's rapid growth continues to aggregate various elements and resources in the southwest, making this young municipality become a modern metropolis step by step. Being the permanent site of the Smart China Expo allows the world to understand Chongqing.

The "grandness of the city" is not contrary to the "temperature of the city". Standing at the beginning of the 14th Five-Year Plan, Chongqing is showing the world "Chongqing-style happy life" with its unique urban characteristics and humanistic spirit.

Section IV

Five Years of Accumulation, Chongqing Adding Color to Life with Intelligence

Just like the growth of human, cities also need to mature in all aspects.

In addition to the common indicators such as economic growth, infrastructure, urban governance, there is another important evaluation dimension in a new era of intelligent industry, which is the level of digital and intelligent development of the city.

Fortunately, Chongqing has not taken too long on the road to intelligent development. At present, Chongqing is ushering in a new wave of technology such as IoT, AI, big data, and cloud computing. Innovative technology has become the driving force for the revolution of Chongqing. With the help of technology, Chongqing's resources have been effectively integrated and utilized, thus stepping into the process of refinement and intelligent management.

When we stand by the Jialing River, and spread the scroll of time in Chongqing in the past five years, we will read a new blend of people, technology and the city, which is actually an answer sheet that adds color to life with intelligence.

"City of Popularity"

An intuitive criterion to determine the value of a city is its popularity. When a city's popularity is strong enough, there is a constant inflow of migrant population, which means that the city is relatively open, and the economy and industry are booming, with enough job opportunities.

"Go to the southwest and open up new markets in Chongqing."

This was the slogan popular with the entrepreneurs a few years ago.

An ideal economic and business environment is more attractive to various factors of production. In June 2020, *2019 China City Business Environment Report* released by the China Media Group showed that Chongqing ranked 5th in the comprehensive evaluation of the business environment of 36 cities in China.[1]

In the image created by domestic mainstream media, Chongqing is the city of industrial construction, the city of new take-off and the city of hope; and in the international perspective, what kind of city is Chongqing? For many overseas media, Chongqing has many different perspectives and a wide range of city labels.

For example, South Korea's *Today Magazine* featured Chongqing as "a key city in the Yangtze River Economic Belt", while industry media such as *Hellenic Shipping News Worldwide, British Rail Freight* website, and *Norway Oslo Transport and Communications Network* all featured extensive coverage of Chongqing's connections to the international land and sea trade corridor and the China-Europe railroad.

In German media such as *Deutsche Presse-Agentur* and *Deutsche Welle*, as well as in South Korea's *Seoul Economy* and the UK's *Automotive Logistics*, the comparative advantages of the China-Europe shuttle train are highlighted because Chongqing is "an important rail hub connecting China and Europe".

In addition to its special geographical location and transportation advantages, Chongqing is also known for its unique and innovative architecture. In addition to the highly recognizable Hongya Cave and other distinctive buildings, such as Raffles City, have attracted a lot of international media attention. "Raffles City Chongqing is an engineering marvel", as reported by media such as *American on Line*, Singapore *Lianhe Zaobao, Saudi Press, The Architect, Portugal Business & Finance News*, and Vietnam's *Labor Daily*.

In 2018, the Smart China Expo was held in Chongqing and settled permanently, attracting a lot of international media coverage. Chongqing reaped new labels such as "technology", "innovation", "intelli-

---

1 Xiao Fuyan, *Chongqing Daily*, "Why Chongqing's Business Environment Can Be Among the Top Five in China", August 21, 2020.

gence" and "smart". *The Associated Press* reported that Chongqing became a media focus not for its beautiful scenery, but for its intelligent advantages.[1]

In the 25 years since Chongqing became a municipality directly under the Central Government, Chongqing has progressed by leaps and bounds, and is constantly updating the world's perception of Chongqing.

In 2010, Chongqing, together with Beijing, Tianjin, Shanghai and Guangzhou, was identified as one of the five major national central cities.

In 2015, the Sino-Singapore Connectivity Project was located in Chongqing.

In 2018, Chongqing began to hold the Smart China Expo, blowing the horn of economic transformation and upgrading.

In 2021, Chongqing, together with Shanghai, Beijing, Guangzhou and Tianjin, took the lead in fostering the construction of international consumer center cities.

In the past five years, Chongqing has been attracting China's top 100 private enterprises and the world's top 500 foreign companies at a faster and faster pace. It is already a pioneer area for the development of inland modern service industries. In 2018, Tencent's southwest regional headquarters and Fosun Group's southwest headquarters settled in Chongqing. In 2019, the western headquarters of YTO Express and the headquarters of Tsinghua Unigroup's DRAM business settled in Chongqing, and Bytedance has settled in Yuzhong District of Chongqing. In 2020, the second headquarters of Qunar, the headquarters of Renren Video, and the national headquarters of Qidi Data Cloud Group were located in Chongqing. In 2021, the second headquarters of the world's top 500 enterprise Amer International Group was located in Chongqing.

Chongqing's entrepreneurial environment and the potential of consumer upgrading are fundamental reasons why major well-known enterprises have settled in Chongqing en masse.

Cities are made up of people. A vibrant and connected, diverse and inclusive, healthy and ecological city is a cradle for stimulating commu-

---

1   Liu Hao, Yin Jia, and Qin Xinjie, *Study on Chongqing's Image in International Media in the Context of "One Belt One Road" Initiative*, July 20, 2021.

nication, innovation and cooperation.

In terms of population indicators, Chongqing is growing again. On May 13, 2021, the Chongqing Municipal Government Information Office held a press conference on the seventh national census in Chongqing, announcing that the resident population of Chongqing was 32.0542 million, an increase of 3.208 million or 11.12% compared with 2010. The growth rate is more than twice the national population growth rate; the inter-provincial inflow population has increased by 1.1516 million in the past 10 years, which is more attractive to talents, and the inflow population has further increased.[1]

Population, labor force and talents, as main market factors, not only drive industrial restructuring, but also are important manifestations of the core competitiveness of urban development. Moreover, beyond the tens of thousands of resident population increment, it may be just the beginning for a growing city like Chongqing.

"City of Technology"

A few years ago, Chongqing was not an ideal destination for the Internet industry.

In recent 5 years, "Smart City" has become another "new business card" of Chongqing.

The concept of smart city is the common product of new technological revolution and new challenges of urban development, and its essence is to empower cities with the means of technology and reshape the development mode of cities. This determines that smart city operators have sufficient potential in technology, data, scenarios and capital.

"Cities are designed to accommodate people". With the development of cities, modern cities have undergone radical changes in terms of urban concept, urban function and urban radiation. Over the years, Chongqing has been thinking and acting in the "big picture", relying on intelligent technology to bring many changes to the city and create a

---

1 *Guangming Net*, "9 Sets of Data, Understand the Changes in Chongqing's Population in 10 years!" , May 13, 2021.

better urban living space.

Over the past five years, intelligence has literally changed urban life and the huge market of "smart cities". If you live in a city and care enough about the infrastructure and municipal news around us, it's probably not hard to find that AI has been integrated into China's urban life as early as in 2017, such as face ID payment in convenience stores.

At the site of 2021 Smart China Expo, Baidu Traffic Brain and Yongchuan Intelligent Transportation case results received wide attention from the industry.

Taking traffic signal adaption as an example, Baidu has optimized and upgraded traffic signals at 29 intersections in key areas of Yongchuan, reusing the constructed outfield sensing equipment, realizing holographic and accurate sensing and analysis and research of urban road traffic status, and creating the first fine-tuned signal timing service in the western region. Real-time adaptive control of 26 intersections and adaptive dynamic filtering control of 1 arterial road have been realized. The average speed of the pilot road has been increased by about 5 km/h and the number of stops has been reduced by about 20%.

Yongchuan realized the first Baidu Map app in the western region to be able to publish the intersection information control information service. At present, the traffic light status and countdown information of the 9 completed signal-controlled optimized intersections can be pushed to Baidu Map in real time to guide the public to adjust the speed in time, help alleviate urban traffic congestion and realize intelligent interaction between roads and vehicles. In addition, in the intelligent bus service, Baidu Map app sets up an independent real-time bus entrance, connects to 28 bus lines in Yongchuan District, and provides real-time information services including bus stops, routes, arrival times to facilitate residents' travel, so that the public can directly share the results of intelligent transportation construction.[1]

With regard to autonomous driving, there are various genres of technology implementation around the world, including the "smart car"

1　Xie Xiaoxi, *Chongqing Daily*, "Baidu Debuts at the 2021 Smart China Expo—Chongqing Yongchuan Intelligent Transportation Project Results Attract Attention", August 26, 2021.

and the "smart road", and it is obvious that the intelligent transportation solution explored by Baidu in cooperation with Yongchuan has become the most cutting-edge exploration of intelligent transportation outside of automobile manufacturers.

In Chongqing, the application transformation of this smart city has long been ubiquitous.

"Smart City"

Across China, as early as five years ago, the smart city strategy has almost become a fad.

The concept of smart city can be traced back to IBM's "Smarter Planet", which was introduced in November 2008 in the theme report *Smart Planet: The Next Generation Leadership Agenda* released by IBM in the U.S. The concept of "Smarter Planet" means to fully apply next-generation information technology to all industries and enhance the digitalization of the whole society.

At that time, the concept was widely recognized, and IBM also paid great attention to China, the world's fastest urbanizing country, and specially moved the presentation to China for another release to promote the concept of "Smarter Planet" and smart cities to the Chinese market.

In 2012, the Ministry of Housing and Urban-Rural Development issued the *Notice on National Smart City Pilot Work*, and smart cities began to be rolled out nationwide. By the end of 2017, more than 500 cities in China have explicitly proposed or been building smart cities.

The collective faith and common practices of cities of all sizes towards the future have kept the smart city construction in China as a whole developing at a high speed.

By March 11, 2021, in *The 14th Five-Year Plan and the Outline of the Long-Term Goals for 2035*, it has been clearly stated: "Promoting the construction of new smart cities by levels and classification, incorporating IoT sensing facilities and communication systems into the unified planning and construction of public infrastructure, and promoting municipal public facilities and buildings, such as IoT applications and intelligent transformation." This represents that the smart city has offi-

cially entered the stage of comprehensive coordination and promotion from the local pilot stage.

Then, from a global perspective, what is the development level of smart cities in China? On December 2, 2021, Deloitte released the report *A Targeted Urban Future: 12 Trends Shaping the Future of Cities in 2030*, which revealed 12 trends for urban life in 2030 through global urban research and observation. After comparative analysis, the report found that China has tended to be a global leader in penetration in areas such as low-carbon smart mobility, automating city operations with AI, and using AI to implement police forecasting.[1]

As a firm explorer of the smart era, Chongqing's smart city construction has made many achievements.

A few years ago, government services were a pain in the neck, but now they are "reporting and approving in seconds". The urban cloud platform has realized the goal of "data run more, and people run less", and constantly updated the scope of "approval online" and "the city's general online office". In terms of the corporate business, in order to constantly optimize the business environment, business start-up approval time has been compressed to within a few tens of seconds. It only takes 6 seconds from application to approval for enterprise project filing. All these are the credit of "cloud government affairs". Weather forecasting, heating, power generation, and water system monitoring are all permeated with intelligence.

Chongqing's emphasis on and conviction to technology has laid a solid foundation for the city's intelligent development. We can also observe that digitalization and intelligence have significantly contributed to Chongqing's urban governance and economic development.

In Chongqing's development plan, smart city will no longer be just the icing on the cake of the city, but the core component of the city's governance capacity and efficiency. The goals of "one-map for comprehensive perception, one-key to know the whole situation, one-stop for innovation and entrepreneurship, and one-screen to enjoy life" are already very clear, and more and more smart applications will allow

---

1 Wang Enbo, *China News Network*, "Interview with Deloitte China Vice Chairman: Smart Cities, Is 'Smart' Enough?", December 4, 2021.

Chongqing residents to enjoy an unprecedented convenience.

All-scenario intelligence is becoming the future city identity. To achieve the overall intelligent development of the city has become the core direction of Chongqing. All departments will be connected to each other to make intelligent judgments and decisions, moving towards a real "city brain". For example, during COVID-19 prevention and control, various application scenarios such as health code, close contact tracking, resumption of work and production and online services have emerged, and application innovation is also driving the renewal and optimization of cloud infrastructure, requiring more business and applications to be born, precipitated, and grown in the cloud from the beginning.

Chongqing, in its meandering process, is alive with the river, nourishing everything and irrigating civilization. This city of mountains and rivers meets the future in a sea of blooming digital flowers.

Part IV

Intelligent Governance: Building an Intelligent Hub for Smart Cities

The intelligent governance of the city seems to be a life with the ability to think, which not only improves the overall operation efficiency of the city from the macro level, but also gives the citizens a delicate and happy experience from the micro level.

The governance of smart cities, with the support of artificial intelligence, is moving from "governance" to "wisdom", and the different focuses of different cities in intelligent governance have formed the distinctive urban characters among smart cities.

Section I

Different Definitions and Reflections of Global Smart Cities in Different Times

Where did the city originate? Standards vary and are still controversial, but it is generally believed that more than 8,000 years ago, Çatalhöyük is considered to be one of the oldest cities in human history.

The city of Çatalhöyük has more than 1,000 houses made of earthen bricks and a population of more than 6,000 people. After the development of the agricultural, industrial and information revolutions, the city after 8,000 years has been completely different, both in terms of quantitative scale and social function.

According to China's seventh census, there are four cities in China with a population of more than 20 million, namely Chongqing, Beijing, Shanghai and Chengdu, and Chongqing ranks first, with a resident population of 32 million.

Today's cities have seen an "explosion" in population and housing compared to the cities in the past, and a greater diversity of urban forms including small cities, large cities, cosmopolitan cities, urban agglomerations, etc. What is more different is the connotation of cities, the physical structure of their operation, and the way they are governed.

The civilization of the city has long undergone a radical change.

Some people may say that the city is the greatest invention of mankind, probably because it has inspired and carried all human civilizations. Globally, the latest consensus on the future of cities is that the civilization of human cities has come to the stage of smart cities based on big data, cloud computing and intelligence.

"Smart" Cities in Different Eras

If a resident of Çatalhöyük, Turkey, had traveled back in time from more than 8,000 years ago to subsequent eras, his understanding of smart cities would have been very different because each era had a different interpretation of the "intelligence" of the city.

Before the advent of the 21st century, there were many things that might have been considered as the "intelligence" of the city by the ancient inhabitants of Çatalhöyük.

For example, if he traveled to Rome in B.C., where sewers were invented and successfully operated, and the city was no longer dirty, he would definitely think: "This is a smart city, which is more civilized and keeps humans away from the possibility of more disease transmission."

If he traveled to London more than 150 years ago, when he first stepped on the subway and was quickly transported to a destination, he would exclaim in his heart: "A smart city should be like this."

Not to mention that 100 years ago in New York, the city's power grid connected the lights and radios of thousands of homes. The residents of Çatalhöyük who came back to the city—looking at the lights of thousands of homes in the dark and listening to the diverse programs on the radio—might feel overwhelmed for a short time, but after getting used to it, he would definitely say: "This is what a smart city should be like."

What is most incomprehensible to the ancient resident of Çatalhöyük is the modern city after the 21st century. By this time, the Internet is superimposed on the power grid. The Internet itself is evolving and has spawned the mobile Internet. The digitalization of industry and digital industrialization brought by the Internet and mobile Internet begin to make things in the city "speak", and then let things and things, people and things can "talk" to each other. At the same time, more and more things not only talk to people, but also help them make better decisions.

This is the new smart city era, where the city is like having a smart brain.

The concept of smart city first appeared in an EU report on *Ranking of European Medium-Sized Cities*, in October 2007, which proposed a

framework of six aspects of smart cities, namely smart economy, smart mobility, smart environment, smart people, smart living and smart management.

In November 2008, IBM officially put forward the concept of "Smarter Planet", and throughout 2009, the concept of "Smarter Planet" and "Smart City" solutions were implemented in more than 50 countries around the world. The term "Smart City" has been established since then, and has been gradually recognized and accepted by the world.

From the content framework of the EU *Ranking of European Medium-Sized Cities Report* [1], it can be seen that the initial smart cities were defined by dividing them into functional modules, such as economy, environment, and residence. This represents a way of definition, but as the economy develops and technology evolves, more and more social functions start to become "smart", and the modules that define smart cities, i.e. the division of dimensions, become more and more detailed, such as smart government, smart education, smart security, and smart manufacturing.

What changes do smart cities bring to citizens' lives?

Although there are different definitions of smart cities from different perspectives and different standards, there are three fundamental perspectives of smart cities, that is, how much smart cities can enhance cities in aspects of interconnection, efficiency and convenience.

Firstly, a smart city should be a connected city, which is subdivided into holistic and comprehensive perception, cross-domain three-dimensional interconnection and executive-level sharing and collaboration. A city's infrastructure is based on new communication technologies. The more interconnected it is, the greater its potential for intelligent development will be.

Secondly, smart cities should be efficient cities, which are subdivided into three aspects: energy-saving and low-carbon, sensitive and convenient, and integrated clusters. The meaning of the existence of cities is to achieve efficient use of social resources and high convenience of

1　Giffinger, Rudolf., Fertner, Christian., Kramar, Hans., Kalasek, Robert., Pichler-Milanovic, Nataša., Meijers, Evert. *Ranking of European Medium-Sized Cities,* Vienna University of Technology, 2007.

urban life with large-scale human settlement. This is always the key to measure the level of urban development.

Thirdly, smart cities should be convenient cities, which are further subdivided into three aspects: wireless ubiquity, integration and personality interaction, creating unlimited possibilities for smart living via big data and communication technologies.

Each era has a different understanding of smart cities, and the same era has different stages of development and focus on the definition of smart cities. But one thing is very clear that the construction of new smart cities is in full swing, and cities have entered into the revolution that amazed the ancients again.

It is sure that this revolution is the most extensive and most changeable one in history.

Smart Thinking and Practice for Global Cities

Since IBM proposed the concept of "smart city" in 2009, countries across the world began to layout the construction of smart cities, and the amount of investment rose year by year.

According to the report *Super Smart City 2.0, Artificial Intelligence Leads the New Way* released by management consulting firm Deloitte in July 2020, the amount of investment in smart cities is expected to peak in the next few years with a compound growth rate (CAGR) of 12.2% in Asia, 8.6% in North America and 9.65% in Europe.

The development paths and means of implementation vary from different countries and their national conditions.

Japan developed the *i-Japan Strategy 2015* in 2009, and Yokohama, a port city of Japan, has been actively practicing this plan. Japan is a country with relatively scarce resources and is frequently hit by natural disasters. Based on its national conditions, Yokohama's smart city construction focuses on "energy conservation".

Specific measures include introducing renewable energy and electric vehicles, implementing smart energy management for homes, buildings and neighborhoods, setting up home energy management systems for citizens, and calling for participation in the implementation of large-

scale experiments on energy-saving actions.

The Netherlands launched the Amsterdam Smart City Approach in 2008. Amsterdam is the largest city in the Netherlands and one of the most important ports in the world, and its construction direction is to "improve the environment, save energy and build sustainable public spaces".

The most noteworthy implementation idea of Amsterdam in the promotion of smart city is "Data-Driven City Management", that is, the construction of a super database, the integration of more than 120,000 data sets of all urban areas, and the open source database. In this way, every citizen or entrepreneur can participate in the city's smart construction based on the use of the data, and enjoy the results of the city's intelligence.

Open source of data is a worthy initiative to inspire the public to carry out the construction of smart cities.

Rio de Janeiro, the carnival capital of Brazil, has a central claim to public safety in building a smart city. To this end, Rio de Janeiro has installed numerous surveillance probes and sensor devices throughout the city, so as to have a clear overview of the city, to react and deploy in a timely manner in case of emergencies, and to guarantee public safety.

New York, the largest city and the largest commercial port in the United States, is also a pioneer in the construction of smart cities. New York's pioneering practice in the field of smart transportation system is worthy of attention, creating a traffic information service system that can track and monitor all the city's traffic status and predict its dynamic changes in a timely manner, and allowing car owners to choose the best driving route according to the system and relevant departments to guide and ease traffic according to the road condition information on the back end.

Singapore launched the "Intelligent Nation 2015" strategy in 2006 and announced its 10-year "Intelligent Nation 2025" plan in 2014. Singapore is converting the entire city into a virtual model, a digital Singapore in a networked world where different city administrations can monitor and collaborate on the city's dynamics.

Singapore's digitalization attempts are based on the application of digital twin technology. This technology has been increasingly used in

city management and smart city construction, which will make the urban governance efficient and accurate.

Different cities have different understanding of smart cities, and different cities have their own focuses on the practice of smart cities, which together build a panoramic view of smart cities in this era. Cities can learn from each other, and continue to improve the construction of smart cities.

China's smart city construction that started in 2009 added an even brighter color to this panorama of smart cities. With a large amount of data accumulated in the mobile Internet era, mature cloud technology, and solid 5G-like infrastructure construction, China is sharing advanced solutions for smart city construction with the world.

Section II

Beijing, Shanghai, Shenzhen and Hangzhou, Intelligent and Innovative Applications in China's Urban Governance

It is an undeniable fact that when looking back through the history and the development of human cities, many of those cities that represent the highest level of human intelligence must have come from China.

For example, in the 7th and 8th centuries, Chang'an (now Xi'an) was known as the "cosmopolitan city", with more than 600,000 permanent residents, including merchants and monks, and the total population was probably more than 1 million. The city covered an area of more than 80 square kilometers, and had communications with more than 300 countries and regions, making it the most prosperous city of its time.

In the 11th and 12th centuries, Bianzhou (now Kaifeng) just covered an area of about 34 square kilometers, but the total population reached about 1.4 million. It was the most populous city in the world at that time, with more than 80,000 artisans and more than 20,000 stores. Every day, the city was crowded with people, horses and wagons. Its trade and commerce were extremely prosperous.

Whereas once benchmark cities were evaluated on the dimensions of population, area and commerce, with the advancement of intelligent technology, intelligence has become another important dimension in evaluating cities.

Similarly, if we were to select the smartest cities in the 21st century, many of them would still come from China.

The Development of China's Smart Cities

As to the meaning and path of smart city construction, there may be different understandings and practices, but China has a broad consensus on the direction of smart city construction and the recognition of its prospects.

China has embarked on the construction of smart cities on a large scale.

According to a report published by Deloitte in 2021, more than 1,000 smart cities have been launched or are under construction world-wide since 2018, with China alone accounting for nearly half of them, and about 500 cities are promoting smart city construction pilots.

Such a vigorous trend of smart city construction cannot be separated from the attention and promotion of governments at all levels.

From 2016 to 2021, China has successively issued policies such as *The 13th Five-Year National Informatization Plan, Application Guide for Key Technologies and Demonstration Projects of IoT and Smart City in 2019, Key Tasks of New-type Urbanization and Urban-Rural Integration Development in 2021,* and *Notice on Promoting the Experience of the Third Batch of National New-type Urbanization Comprehensive Pilot.* China has clarified the strategic position of smart cities as its urbanization development and sustainable urban development programs, as well as the task of "promoting the construction of smart cities", stimulating the construction of smart cities around the country. This has a great positive effect on the development of smart cities and related industries.

It took China about 13 years from the proposal of the concept, city pilot, promotion and implementation to mature ideas and successful cases.

At the end of 2012, China's Ministry of Housing and Construction launched the first batch of national smart city pilots, including a total of 90 pilot cities (districts and towns) in Beijing Dongcheng District, Beijing Chaoyang District, Shanghai Pudong New Area, Hangzhou Shangcheng District, Shenzhen Pingshan New Area, Chongqing Nan'an District, and Chongqing Liangjiang New District.

In 2014, the National Development and Reform Commission, the

Ministry of Industry and Information Technology, the Ministry of Science and Technology, the Ministry of Public Security, the Ministry of Finance, the Ministry of Land and Resources, the Ministry of Housing and Urban-Rural Development, and the Ministry of Transport issued the *Guidance on Promoting the Healthy Development of Smart Cities*, proposing the construction of a number of distinctive smart cities by 2020.

In the same year, 26 departments jointly set up an inter-ministerial coordination working group to promote the healthy development of smart cities, which is at the level of institutional setting to ensure the efficient and smooth development of smart cities.

In 2016, the State Council issued the *Notice of the 13th Five-Year National Informatization Plan*, formally proposing the action of new smart city construction, and specifying the leading units as the National Development and Reform Commission and the Office of Network Security and Informatization Commission. This formally determined the development of new smart cities, confirmed the construction of new smart cities as a national project, and promoted the construction of smart cities into the fast lane.

For the construction of smart cities, China has developed a complete industry chain with corresponding representative enterprises.

According to the classification of iResearch, the smart city industry map can be divided into 5 layers.

The first layer is the top layer of smart city design, including Huawei, Taiji Group, Digital China, Inspur Group, CETC and other companies that provide overall solutions for smart city construction.

The second layer is the sensing and communication layer, which contains companies such as China Mobile, China Telecom, China Unicom, ZTE, and Datang Telecom. In other words, these companies mainly focus on the collection and transmission of data.

The third layer is the platform and infrastructure layer, which includes Alibaba Cloud, Ping An Technology, H3C, Sugon, and Taiji Group.

The fourth layer is the urban computing layer: it contains Jingdong Digital Technology, Alibaba Cloud, Tencent Cloud, Baidu Cloud, iFlytek and Ping An Technology. Simply put, this is the layer that provides computing power and algorithms.

The fifth layer is the application layer of subdivided scenarios, which is also the layer most easily perceived by residents. For example, the construction of smart traffic cannot be separated from the monitoring equipment of Hikvision, and similar application layer enterprises include CTFO PCITECH, New CSP, and e-Hualu.

In addition, there are also some companies specializing in urban security such as 360, Sangfor, AsiaInfo, DBAPP Security, and HISPE-CIEL throughout the upstream and downstream of the industry.

At present, the main construction participants of domestic smart cities are still dominated by large technology enterprises, including Jingdong Digital Technology, Alibaba Cloud, Tencent Cloud, Huawei, Ping An Smart City and Inspur. They have both leading solutions and technologies, thus can fully cover different levels of the smart city industry map.

The construction of a smart city is a huge project that requires scientific research, and the application of technology requires the broad participation of the government, commercial companies, research institutions and the public.

Currently, the construction of domestic smart cities follows a market-oriented approach: government input is the main body, supplemented by cooperation with strong commercial companies; strategic planning and top-level design are carried out by commercial companies, research institutions and think tanks to ensure the right direction and the effectiveness of practice; application development is a joint effort between the government and commercial companies.

In short, the wide participation of all sectors of the society is driving the development of smart cities in China.

Intelligent and Innovative Applications in Beijing, Shanghai, Shenzhen and Hangzhou

Each city in China has different basic conditions and resource endowments, and the path and focus of developing smart cities also differ. Because of the differences among cities, a variety of smart city development samples can be found in China. Through the differentiated ex-

ploration of different city samples, China is likely to provide a reference for the global smart city construction that is closer to the ideal goal.

Beijing is committed to building a global benchmark for new smart cities.

On the one hand, Beijing continues to strengthen the "sensing capability" of the city, promoting the construction of smart towers and other sensing bases networking, and realizing the comprehensive carrying of multiple devices and sensors; actively promotes the construction of gigabit broadband access network and accelerates the deployment of IPv6-based next-generation Internet to clear the obstacles in data and signal transmission for IoT, data transmission and autonomous driving.

On the other hand, in order to make urban governance more effective, Beijing is building a "one-network management" application system, a support system for urban operation management and decision making with the core of accurate urban population management, economic activity monitoring, urban operation awareness and comprehensive urban management and law enforcement. On the basis of such a system, decision-making will be made based on more comprehensive and timely information, and urban governance will be more efficient and accurate, realizing data-driven intelligence.

Different cities have similar concepts of smart city governance. Beijing applies "one-network management" while Shanghai adopts "one-network office".

The "one-network office" is the brand of government services in Shanghai. According to the statistics of October 2021, Shanghai has accessed more than 3,400 items on the "one-network office" platform, with an average daily volume of 160,000 items.[1] As more and more management departments connect their public services to the "one-network office" platform, the volumes of data will grow, behind it will be a more convenient and livable city life, and a more open and efficient business environment.

There are many humanized designs in Shanghai's intelligent governance. The smart elevator platform is a good example. With the help

---

1 Zhou Di, *The Paper*, "Shanghai: 'One-Network Office' Handles 160,000 Items Per Day, Continuing to Deepen the Application of Data Governance", November 19, 2021.

of remote monitoring automatic identification, the average time for emergency rescue of elevator trapped people in Shanghai is now shortened to 12 minutes.[1] In Changning District, Shanghai, a smart water meter installed in the home of an elderly person living alone will automatically notify the community worker if the reading falls below 0.01 cubic meters within 12 hours, so that someone can be sent to the home for inspection in the first place.[2] With these small innovative applications, Shanghai is able to solve a subtle problem more intelligently and build a safer and more considerate urban environment step by step.

Now, the Shenzhen Municipal Government, together with Huawei, has built the Intelligent Operations Center (IOC), which is the "brain" of the "digital government". This is a smart city hub that can see, use and think, connecting 42 systems, more than 100 types of data and more than 280,000 video channels, building up Shenzhen's "1+12+N" integrated command system, and forming a solid and powerful government data "base".[3]

The innovation and exploration of smart city in Shenzhen is marked by the joint efforts of government and enterprises. Every enterprise can find its own role in the construction of smart Shenzhen. In Shenzhen, "Yue Shengshi" (a service mini app) built by Tencent and "iShenzhen" built by Ping An Smart City have realized government services from "offline" to "online" and "online" to "handheld". The multi-dimensional water model system jointly built by Huawei, Shenzhen Water Bureau and China Institute of Water Resources and Hydropower Research has achieved the expected results of maintaining the ecological health of rivers and lakes on sunny days, reducing pollution in the first rain, alleviating waterlogging disaster in heavy rainy days, and ensuring the safety of urban flood control in rainstorm weather. The world's first Construction Digitalization Management Center (CDMC) created by

---

1 Liu Hao, *China Consumer News*, "2021 'Shanghai Standard' Released: Elevator Emergency Rescue Time Shortened to 12 Minutes", October 25, 2021.

2 Li Shu, *Chongqing Daily*, "Less Than 0.01 Cubic Meters in 12 Hours to Alarm, How the Smart Water Meter Reads the Life of the Elderly Living Alone?", December 19, 2020.

3 Wang Xinguan, *Shenzhen Evening News*, "Deciphering the 'Smart Code' of a City of Innovation, Huawei Helps Build a Smart City in Shenzhen", May 21, 2020.

Huawei and Shenzhen Metro innovatively realized the centralized management and control of more than 400 subway construction sites, comprehensively improving the Information and Communications Technology (ICT)-based management of urban rail transit construction.[1]

As early as April 2016, Hangzhou started an AI exploration called "city brain", but initially, it was only a breakthrough in the field of transportation, using big data to improve urban transportation. It was then that Hangzhou took a step forward from "reducing traffic congestion" to "city governance" and built an intelligent platform covering 11 areas such as public transportation, city management, health and grassroots governance, which includes 38 application scenarios and 366 business items. It can find parking places and check the weather, as well as provide health codes, public opinion DirectGet and 12345 livelihood platform[2], making Hangzhou a city of digital system governance.

In 2022, Hangzhou City Brain ushers in the 2.0 version, taking the community as the entry point. The first batch focuses on areas such as housing, transportation, elderly people and children, promoting the construction of major projects such as housing, transportation, "one-elder and one-child" caring service in the future community.

In City Brain 2.0, seven sub-scenarios of smart planning, smart traffic management, smart bus, smart subway, smart logistics, smart parking, and smart slow traffic will be built in the smart traffic scenarios with precipitation and accumulation, creating smooth, safe, convenient, efficient and green urban traffic.

1   Zhou Yumeng, *Shenzhen Special Zone Daily*, "Shenzhen Smart City Construction Delivers a Bright Answer Sheet as 2020 Comes to a Close", December 28, 2020.

2   Huang Ping, *Economic Daily*, "Install a 'Brain' for the City—Survey on Hangzhou's Smart City Construction", March 23, 2021.

Section III

Interpreting Smart Chongqing, the Surge in Intelligent Urban Governance

Chongqing citizens who are concerned about the 2019 Smart China Expo must remember two phrases: "smart manufacturing town" and "smart city".

On August 27, 2019, at the opening ceremony of the 2019 Smart China Expo, Chongqing proposed to focus on building a "smart manufacturing town" and a "smart city", and to create a smart era and share smart achievements with solid efforts.

These two phrases are the label of Chongqing to build a smart city, as well as the vision of Chongqing towards the smart era, focusing on the economic development of Chongqing, taking into account the living experience of Chongqing, and echoing the theme of "Empowering Economy, Enriching Life" of the Smart China Expo.

Taking "smart manufacturing town" and "smart city" as the vision and goals of Chongqing smart city construction reflects the rapid progress of Chongqing's intelligent capacity and governance, as well as the confidence and determination of Chongqing in the smart era.

The History of Smart Chongqing

Looking back at the history, it can be said that Chongqing's smart city construction started in 2000, and accelerated in 2018.

In 2000, Chongqing was approved by the Ministry of Construction (now Ministry of Housing and Urban-Rural Development) as a demonstration city of "digital city", and the digital Chongqing construction at that time was more of a digital record of the city at the physical level,

that is, connecting the city to the Internet.

For example, the completed database contains basic geographic information database, underground space database, toponymic address database and comprehensive pipe network database. The completed software platform contains basic geographic information platform, governmental geographic information platform and social service geographic information platform.

"Geography" is the keyword of digital Chongqing construction, and the next keyword is "IoT".

In terms of the IoT, Chongqing has successively compiled and issued *The 12th Five-Year Plan for the Development of the IoT Industry*, and the *Opinions of Chongqing Municipal People's Government on Accelerating the Development of the IoT*.

The shift from "geography" to "IoT" reveals the significant progress of Chongqing's digital construction, that is, in the first decade of the 21st century, Chongqing has completed the datafication of urban geography and space, and gradually expanded the scope of datafication to all aspects of urban governance and urban life. "IoT", therefore, has become the keyword to Internet of Everything (IoE) and precipitate data to achieve more dimensional intelligent exploration.

Besides, the shift from "geography" to "IoT" is a process to realize the change from "digitalization of two-dimensional Chongqing" to "digitalization of three-dimensional Chongqing", which relies on the development of sensor equipment. With qualitative changes in data dimensions and volumes, cloud computing becomes more and more important. Therefore, Chongqing began to build the Liangjiang International Cloud Technology Center, and established the Chongqing Cloud Computing Association.

At the end of 2012, the Ministry of Housing and Urban-Rural Development launched the first batch of national smart city pilots, and Chongqing's Nan'an District and Liangjiang New District were included on the list. The concept of smart city gradually popularized throughout the country.

In 2018, two major events took place in Chongqing, which opened a new phase of smart Chongqing construction. One is the successful holding of the 2018 Smart China Expo in Chongqing, which has had

a profound impact on the construction of Chongqing's smart city, as reflected in the convergence of wealth and wisdom, enterprises, industries, consensus and other aspects. The other is the establishment of Chongqing Big Data Application and Development Administration.

As one of the first eight big data administrations established in China, Chongqing Big Data Application and Development Administration is responsible for big data-related duties, and the construction of smart cities as well. The significance of the establishment of this administration is not only that there is a specialized organization responsible for the construction of smart city, but also that together with the "cloud officer system" launched by Chongqing later, it sets the new phase of building a smart Chongqing.

In 2019, Chongqing issued the *Implementation Plan of the "Cloud Officer System" in Chongqing*. It is a structure system that the main leaders of the Municipal Government act as "general cloud officer", six systems including politics, law and transportation set up "system cloud officer", and the main heads of all departments of the districts, counties and municipal governments serve as the "cloud chief" of each unit.

Smart city construction depends on both hard power such as infrastructure and hardware as well as soft power including application ideas and data. Data is the key, which is known as the driving force and resources of the smart era, and is as important as electricity to the age of electricity. The history of Chongqing smart city construction reviewed earlier is essentially a history of data collection, computing and application, and the difference lies in the data collection capacity, computing capacity, transmission speed and application scenarios.

The foundation of smart city construction is data sharing, and data sharing is divided into two steps: collecting and sharing. In short, at the data level, the "cloud officer system" ensures that the data from all departments and counties are uploaded to the cloud and meet certain standards and conditions, which is a solution to the problem of availability and quality of data. The leadership of Big Data Bureau can well coordinate the application of data and solve the problem of data interconnection and sharing.

The collection, uploading, sharing and application of data correspond to the mechanism function of "cloud officer system" and the

institutional responsibilities of the Big Data Bureau. In 2021, it is reported that Chongqing has realized the sharing, aggregation and opening of government data, which are 4,055, 3,309 and 1,310 categories, respectively. A total of 14.2 billion pieces of data have been called, and the invocation and application of each piece of data[1] has promoted the intelligence level of Chongqing's governance logic and lifestyle.

The Construction System of the Famous Smart city

The "cloud officer system" and the Big Data Bureau have solved the data-related pain points and choke points from the institutional level, so that the "blood" of smart city construction can flow freely, and the "driving force" can be released to the fullest.

The empowerment of the key element of data in the construction of smart city has laid a solid foundation for the rapid advancement of smart city governance in Chongqing in recent years. Data is a very critical point in the construction of a smart city, and the intelligent development of the entire city also depends on overall planning and comprehensive efforts.

In April 2019, Chongqing adopted *Chongqing New Smart City Construction Program (2019—2022)*, proposing to promote the construction of Chongqing new smart city with "135" overall architecture, which is to build a city intelligent hub composed of digital Chongqing cloud platform, urban big data resource center and smart city comprehensive service platform, to consolidate the three major support systems: the new generation of information infrastructure system, standard evaluation system and network security system, and develop five types of smart innovation applications: livelihood services, urban governance, government management, industrial integration, ecological livability.[2]

In addition, Chongqing is accelerating the construction of "8611"

---

1 Cui Li, *Chongqing Daily*, "Focus on Solving 'Pain Point', 'Choke Points', and 'Difficult Points', and Connect the Underlying Data to Enhance the Construction of Chongqing's Smart City", February 24, 2022.
2 *Chongqing Daily*, "Chongqing Municipal Government Held the 45th Executive Meeting", April 10, 2019.

integrated scenario construction system. The construction path of "smart city" is more refined and the pace is accelerating.

"8" refers to 8 basic capacities, which means accelerating the construction of the national integrated computing network Chengdu-Chongqing national hub node, "3,000 trillion" city, AIoT infrastructure, urban information model platform, "mountain city chain", upgrading urban big data resource center, digital Chongqing cloud platform, promoting the Sino-Singapore international data channel, and improving the infrastructure capacity of digital, network and computing scenarios.

Different from the institutional empowerment on data by the "cloud officer system" and the Big Data Bureau, the eight basic capacities are actually empowering data from the technical level to improve the speed of data transmission, storage and computing capabilities, and enrich the data collection dimension and sharing scenarios.

"6" refers to 6 support systems, the integration of "data directory, data standards, logical architecture, system interfaces, business processes and capability components" can continuously improve the data support capability, system integration capability and business collaboration capability, consolidating the common support of cross-scenarios.

The first "1" refers to over 10 cross-platforms. "Yukang Code" is one of the "cross-platforms", which realizes cross-tiered, cross-regional, cross-systematic, cross-departmental and cross-business collaborative management and services.

"Yukang Code" is a typical innovation born from the combination of big data and cloud computing in Chongqing's smart city construction under the special demand of COVID-19 prevention and control, which can provide citizens with the query function of COVID-19 prevention-related information, such as nucleic acid test, vaccination, travel status, etc. The function of "Yukang Code" depends on two aspects: one is the data collection from health, public security, industry and information, transportation and other departments, and the other is the sharing and application of these data.

In addition to "Yukang Code", the "cross-platforms" also include "Yukuaiban", which provides government services for the public. According to Chongqing's open data in the second half of 2021, the matters that could be handled by "Yukuaiban" have reached 1,875, and the

number of users has exceeded 21 million. In addition, "Yukuaizheng" has been applied in 38 units on a pilot basis, and the provincial government's online service capability assessment has ranked in the top 10. "Yukuairong" has more than 260,000 users and financed 37.1 billion yuan for small-and-medium-sized enterprises. The capability of smart city has been expanded from serving the general public to government affairs groups and small-and-medium-sized enterprises.

The last "1" refers to 100 typical applications, that is, to create the whole scenario set of "housing, education and employment, tourism, entertainment and shopping" oriented by application. "Housing" will focus on "future community", including smart medical care, smart nursing homes and smart firefighting. "Education and employment" will focus on creating scenarios such as smart education and smart employment. "Tourism" will focus on scenarios such as smart transportation and smart scenic spots. "Entertainment" will focus on scenarios such as digital entertainment experience and smart sports. "Shopping" will focus on scenarios such as smart business district, smart finance, and smart logistics.[1]

If the increase of cross-platforms like "Yukuaiban" is making Chongqing's urban governance more intelligent, the popularity of smart applications related to "housing, education and employment, tourism, entertainment and shopping" is adding the color of intelligence to urban life.

If there is battery failure and low power of new energy vehicles, the back-end power battery risk real-time monitoring and warning big data platform can remind the driver in real time. We can buy attraction tickets and book hotels on the app, and wear VR glasses to travel through time and space and visit the whole Chongqing. Walking into the mall and clicking on the touch screen, we can see the introduction information of all stores, and there are robots to take us to the corresponding stores.

Communities are getting smarter and smarter. Some communities have combined face recognition technology with the community access

---

1 *Chongqing Daily*, "Chongqing Takes the Lead in Smart City Construction in China", August 18, 2018.

control function. The door immediately opens when you scan your face. The access control system is also linked with the system of public security authorities to achieve real-time supervision and real-time warning of the flow of people and vehicles entering and leaving the community. Some communities have greatly reduced the occurrence of throwing objects from the high by installing smart monitoring systems. Some communities have achieved 24-hour health management and precise services for the elderly, disabled, children and other groups by introducing a comprehensive health IoT system...

The factor of intelligence, under the overall planning of the construction of a famous smart city, is seeping into every cell of Chongqing along with data, platforms and applications. A new and smart city is coming at an accelerated pace.

Section IV

Intelligent Urban Governance, the Future of Development Space

Boosted by the development requirements of the new smart city construction, cities across the country have made significant improvements in their governance capabilities.

Shenzhen's smart city construction takes the overall construction of "one-system for sensing network, one-identification for all public services, one-click to obtain all data, one-integrated urban operation management system for synergy, one-service platform for innovation and entrepreneurship, and one-integrated citizen service platform to enjoy smart life" as the goal.[1] The effective connection and intelligent interaction of data break the barriers between different departments and fields, and solve the problem of governance fragmentation.

The document on accelerating the construction of a smart city in Shanghai details the specific requirements for accelerating the urban operation of "one-network unified-management" and the integrated construction of an urban operation system, closely following the goal of "one-screen to view the world and one-network to manage the whole city", and accelerating the formation of a cross-departmental, cross-tiered and cross-regional collaborative operation system.

In Chongqing's planning, the new smart city operation and management center will be fully completed in 2025. By then, the urban operation will be implemented with "one-network management", "one-network office" for government services, "one-network scheduling" for emergency management, "one-network governance" for grassroots

---

1 Zhou Yumeng, *Shenzhen Special Zone Newspaper*, "Shenzhen's 'Six-Ones' Concept Leads Smart City Construction", October 14, 2020.

affairs, and to improve the overall level of digital governance.[1]

With the help of intelligent means such as cutting-edge technology and advanced equipment, intelligent sensing, intelligent identification, automatic warning and other technologies are widely applied and are becoming a powerful support for the fine and high-quality urban governance.

Intelligent urban governance has become an important part of meeting the people's growing need for a better life.

The Evolution of Cities from "Governance" to "Smart Governance"

The challenge posed by the growing size of cities is the increasing complexity of urban governance.

The issues of community massification, atomization of residents, diversification of demands and sustainability have become the brand-new challenges for urban governance.

Essentially, the city is a high degree of aggregation and coordination of people and resources in a limited space. With the advent of the data era and the establishment of smart cities, some changes are taking place in the focus of city governance.

If striving to create a favorable environment for people to work, live, enjoy and travel is urban governance 1.0, then focusing on people's sense of security, gain and happiness is urban governance 2.0. The rise of urban governance from "construction of things" to "service for people" also leads to the evolution of urban governance from "governance" to "smart governance".

Big data and cloud computing in the age of intelligence have opened a new chapter for urban governance, and the intelligent technology begins to add "intelligent wings" to urban governance.

The concept of smart governance relies on intelligent technology and bring together the wisdom of all people to implement refined gov-

1 Xiang Jumei, *Chongqing Daily*, "Chongqing will Build a National First-Class Data Gathering Land and Utilization Highland", January 5, 2022.

ernance of the city. With intelligent technology as the "brain" of the city, governance will be "smarter". Key technologies such as IoT, geographic information technology, network communication technology, big data, cloud computing and social computing play a crucial role in data collection, transmission and processing.

There are numerous examples of intelligent technologies used in urban governance. Smart security protects our safety, smart justice helps maintain social fairness and justice, and smart elderly-care service copes with the arrival of a severe aging society. In addition, with the unified scheduling of smart city construction and the strong support of intelligent technologies, the city's environmental management, public health, traffic and other aspects of urban management has achieved lower cost, higher efficiency and better experience.

In the specific operation, the governance can accurately, easily and timely locate the problem and achieve targeted and fine governance.

In the past, if the street lamp was broken, the manhole cover was lost, and the road was flooded, it was difficult to solve the problem in the first time only by the staff inspection or the citizens' clues. Now, sensors and other big data devices have become the brain, eyes, ears, mouth and hands of a city. As long as the smart city operation and management center is connected to carry out real-time supervision and timely release the abnormal situation detected, the staff can arrive at the site to deal with it within half an hour.

In community governance, smart governance promotes the urban operation of "one-network management", and "one-network office" for government services. It has been very effective in enhancing the effectiveness of policy advocacy, public communication and convenience services, and has truly enabled the public to realize the goal of "data runs more, and citizens run less".

Space for the Future of Urban Governance

According to the vision of smart city construction, the goal of cultivating fine urban governance capacity is to achieve overall governance based on the realization that the decision-making, execution, monitoring

and evaluation of urban governance are all evidence and data-based.

There is a popular saying about the so-called data-based principle: if the public data "runs fast" and the efficiency of government governance will be increased. Since government departments hold 80% of the data resources, opening government data is conducive to breaking down departmental information barriers, releasing data value, and improving governance and public services.

In the digital era, data has become the new factor of production.

Ni Guangnan, an academician of the Chinese Academy of Engineering, expressed in his speech at the 2021 Smart China Expo that at present, data has become a basic strategic resource in China, and an important factor of production and productivity, and the support of data is of great significance in empowering the development of emerging technologies, promoting the digital transformation of industries, enhancing social governance, and safeguarding national cyber security.

Where is the future of intelligent urban governance? Deeply cultivating data governance is a critical key to this question.

As a major digital economy and data governance country, China carries out data governance which is of great significance to drive the innovative development of digital economy, comprehensively improve the effectiveness of government governance, and empower public services and social governance. According to the latest *V1 Global Big Data Spending Guide 2021* released by market research firm IDC, the global big data market spending is expected to reach about US$298.30 billion in 2024, achieving a CAGR of about 10.4% over the five-year forecast period (2020—2024). China's big data market is developing rapidly and leading the world in growth rate, with a five-year CAGR of about 19.7%, and the total market is expected to exceed US$20 billion in 2024.

Empowering urban governance and improving urban effectiveness with data governance is a common desire of operators, government and enterprises in recent years. However, the road to data governance construction is still full of challenges, such as data silos, data risks and how to achieve effective regulation.

How to manage data? How to share data?

In terms of data silos, the government itself needs to build an

integrated digital platform for government affairs, linking all departments and tiers within the government to achieve smooth information exchange, breaking the limitations of "data silos" and "data towers", realizing an "integrated" office model with multi-tiered and multi-departmental collaboration and synergy, avoiding complicated and cumbersome government procedures, and achieving the coordination of government resources.

In terms of data risk, He Xinfei, vice president of QI-ANXIN Technology Group, shared at the 2021 Smart China Expo Forum that they hope to have an endogenous definition of data security, which can enable data governance while achieving the following functions: identify the systems and networks where sensitive and important data are distributed and the locations where they are stored, and realize the monitoring and timely discovery of data as they flow.

As to how to achieve effective regulation, some cities have taken the lead in issuing relevant regulations and implementing legislation in order to protect data ownership, use rights, privacy rights and other rights from infringement. For example, Shenzhen announced the "Shenzhen Special Economic Zone Data Regulations", which became the first local basic and comprehensive legislation in the field of data. Chongqing adopted the "Chongqing Data Regulations", which has come into force on July 1, 2022. It establishes and improves the data processing rules and data security system in the context of data security in Chongqing.

National Urban Data Governance Is Developing in Depth

Enjoying convenience and efficiency in life is the most practical experience brought to the people by the national data governance project.

The 2022 government work report proposes that we will work to build a digital government and promote the sharing of government data, continue to simplify various certification requirements and provide more government services on a cross-provincial basis, basically realize mutual nationwide recognition of electronic licenses and certificates to facilitate enterprises with trans-regional operations, and provide people

with greater access to important services outside of their home jurisdiction. We will facilitate integrated access to government services, and introduce simpler procedures for the registration of immovable property and motor vehicle inspections.[1]

Over the past two years since its launch, the national government service platform has brought together the government service affairs of 31 provinces (autonomous regions and municipalities directly under the Central Government) and Xinjiang Production and Construction Corps as well as 46 State Council departments, providing services covering electronic licenses and multiple fields such as education, disability assistance, justice and civil affairs.

The continued upgrade of data governance capabilities in major cities has brought many conveniences to people's lives.

In 2018, under the new requirement of "new smart city", Shenzhen put forward the development goal of "one-system for sensing network, one-identification for all public services, one-click to obtain all data, one-integrated urban operation management system for synergy, one-service platform for innovation and entrepreneurship, and one-integrated citizen service platform to enjoy smart life". In the same year, Shenzhen took the lead in launching the "approval within seconds" reform of government services to achieve "online-accessible-affordable, once-for-all handling".[2]

In 2019, Shanghai continued to enrich its platform by promoting the "one-network office" government services. By the end of 2021, 3,458 government service affairs have been accessed, with a cumulative service of over 13.66 billion people and nearly 200 million cases. In 2021, the online handling rate of "one-network office" reached 77.03%, the full online handling rate reached 69.30%, and the peak monthly activity of the "apply-as-you-go" service exceeded 18.58 million.[3]

---

1 Bo Chendi, Shen Yaxin, *Peoples Network*, "The 2022 Government Work Report: Promoting the Sharing of Government Data and Providing More Government Services on a Cross-Provincial Basis", March 5, 2022.

2 *China Business Network*, "One Law, One Network and One Line, Shenzhen Has These Experiences in Data Governance", September 28, 2021.

3 Wu Di, *Liberation Daily*, "'One-Network Office' Will Promote 'No Application to Handle' Service", January 5, 2022.

Chongqing is unwilling to lag behind either. In December 2021, Chongqing Municipal Government issued *The 14th Five-Year Plan of Chongqing Data Governance (2021—2025)*. It clearly proposed that by 2025, the city's big data resource center, data mapping and urban information model will be fully completed. The data aggregation rate will be no less than 90%, the number of government data sharing will be no less than 20,000, and the number of public data opening will be no less than 5,000. The accuracy, timeliness and availability of data will continue to improve, and the level of data sharing and opening will be at the forefront across China.[1]

It is worth mentioning that at the 2021 Smart China Expo, Jingdong Technology Group exhibited the first integrated management system of intelligent environment of industrial park in southwest China built in Chongqing with Xiong'an New District Block Data Platform and Nantong Municipal Governance Modernization Command Center. The system, supported by the powerful backend storage of Jingdong Cloud and the efficient spatio-temporal big data governance capability of Jingdong Smart City Operating System, brings together and shares the environmental protection information of 334 enterprises in Chongqing Economic Development Zone with a single screen for operation. Through big data analysis and research, the AI model can monitor and warn about the excess coefficients of wastewater and smoke, timely find the high pollution, high-energy consumption and high-water consumption enterprises, and scientifically assign personnel to the site for disposal, using technology to guide the park to save energy, reduce emissions and avoid environmental accidents.[2]

The national urban data governance is developing in depth, and its application scenarios are blossoming. The city's dream of the future is also being realized step by step.

1 Liang Haonan, *CQNEWS*, "The '14th Five-Year' Plan of Data Governance Is Coming! By 2025, Chongqing Will Fully Build a New Smart City Operation and Management Center", January 5, 2022.

2 Xiang Jumei, *Chongqing Daily*, "Become the First to See These 'Black Technologies' at the Smart China Expo", August 21, 2021.

Part V

Chongqing:
An Exploration of a
Brand-New Smart City

The development of a city, like the growth of a person, often moves forward in exciting years, and occasionally hesitates at confusing points in time. How does a heavy industrial city in the inland create a miracle of growth for consecutive years, and how does it switch engines at the crossroads of the times? Chongqing's metamorphic journey is clearly worth thinking.

Where will the future of smart cities go? Will they continue to rely on the evolution of technology? Or will they depend on the self-formed open innovation ecology? Chongqing also has its own thoughts and choices.

Section I

Industrial Upgrading: From "Manufacturing Town" to "Smart Manufacturing Town"

Over the past five years, from a serious study of the summit to a careful look at the times, and then looking back at the city, a new perplexity has arisen unconsciously, which is also the perplexity to trace the origin. Why can Chongqing quickly form a competitive industrial cluster when an era of intelligent industry is rumbling so clearly, with almost no time lag and information gap around the world?

Obviously, this question cannot be answered in a few words because of the numerous dimensions and wide fields of variables involved. Moreover, the process of seeking for an answer is easily pushed by the questioner to the dead end of realistic details and to provide a symbolic answer with figurative details that can hardly support the whole, thus ignoring the key history of this manufacturing city.

As one question remains unanswered, other questions pop up: Can the opportunities Chongqing has grasped in the intelligent industry trend be replicated in other similar or heterogeneous cities? What are the key strategic points and tactical methodology in the process of replication?

These questions are even more acute than the previous one because the transformation of Chongqing is still a work in progress. Riding on a fast lane and still exploring its future, it is difficult to summarize the route and experience before reaching the desired destination, let alone leave a map that can be traced for future generations.

For Chongqing, the process of trying to answer the above questions is an exploration in an era and seeking the testimony of an era as well.

Five Consecutive Years of Industrial Upgrading and Growth Miracle

In the past five years, Chongqing's GDP has increased by nearly 1 trillion yuan, with an average annual growth rate of 6.7%. In 2021, it reached 2.79 trillion yuan, and per capita regional GDP reached 87,000 yuan, which was higher than the national average. The supply-side structural reform in Chongqing continued to deepen, the pace of industrial transformation and upgrading was accelerated, and pillar industries such as automobiles and electronic information continued to grow. Chongqing's innovation-driven development led by big data intelligence has been further advanced, and breakthroughs have been made in the construction of "smart manufacturing town" and "smart city". Chongqing has successfully held four Smart China Expos, and the added value of digital economy accounted for 27.2% of the regional GDP ...[1]

Given the volume of Chongqing's GDP, it is a miracle to maintain such a high growth rate.

Chongqing has been an old industrial bases. In recent years, outperformance has been widely interpreted as Chongqing has made better progress in economic structural transformation around industrial restructuring, promoted supply-side structural reform, and increased innovation and openness.

Yi Xiaoguang, president of Chongqing Comprehensive Economic Research Institute, said that Chongqing has been vigorously promoting product restructuring and industrial restructuring in the past few years. In the past, the product structure of the automotive industry was biased towards the middle-and-low end, and now it is transforming to the middle-and-high end. In addition to the laptop industry, other industries such as LCD panels are also emerging in the electronic information industry. These industries are driving Chongqing's economy out of the downturn. In 2021, the value of automobile and motorcycle electronics industry, two pillar industries in Chongqing, increased by 11.3% and 17.3%, respectively, of which the automobile industry increased by

---

1 Luo Jingwen, *Chongqing Daily*, "Keep in Mind, Forge Ahead and Record 5-Year Achievements in the History of Chongqing", May 28, 2022.

12.6%.[1]

Chen Yaoxiang, vice president and secretary-general of China Regional Economic Association, believes that the industrial restructuring of the northeastern provinces is still in the throes of change. Part of the industries in Liaoning are "the pillars of a great power" with high-end equipment, and they are impossible to expand on a large scale like other industries, so they are limited to some extent.[2]

Chongqing's "Intelligent Transformation" Journey

So, what is the code of the rapid economic growth of Chongqing?

According to Pan Helin, co-director of the Digital Economy and Financial Innovation Research Center at Zhejiang University's International Business School, there are three reasons. First, Chongqing gets rid of the GDP dependence on industries with overcapacity and real estate, and focuses on developing strategic emerging industries and real economy. The industrial growth has become the "anchor" of Chongqing's GDP. Second, Chongqing does not blindly follow the eastern coastal areas. Its innovative processing trade model makes industrial transfer and upgrading no longer a rootless "duckweed" and becomes the main force of Chongqing's GDP growth. In addition, the social system, especially the innovation of property rights transactions (such as the land index system and the promotion of urban-rural integration), has brought new momentum to Chongqing's GDP growth.[3]

In recent years, Chongqing does not focus its resources on industries with overcapacity and the price of housing has been relatively well controlled. Chongqing focuses its resources and efforts on strategic emerging industries such as automobiles, electronic core components,

1   Chongqing Municipal Bureau of Statistics, Chongqing Survey Team of the National Bureau of Statistics, *Chongqing Economic Operation in 2021*, January 19, 2022.

2   Li Xiuzhong, *China Business Network*, "After Sichuan, Chongqing's GDP Also Expected to Exceed Liaoning, Southwest Economy Fully Overtaking That of the Northeast?", January 22, 2022.

3   Pan Helin, *Beijing News*, "What makes Chongqing's GDP No.1 in China?", October 28, 2016.

IoT, high-end intelligent equipment and new materials, and the real economy, making the industrial growth become the main contributor to GDP growth. Chongqing's practice of guiding finance to support strategic emerging industries and the real economy through the establishment of funds has become a catalyst for changing the city's GDP gene.

Since China's reform and opening up, especially after its accession to the WTO, playing the role of "world factory" and developing processing trade has become an important force for the economic growth of the eastern coastal cities. However, most of the traditional domestic processing trade industries have a short industrial chain. Some industries have to purchase raw materials and components from overseas and sell them overseas after processing. This kind of manufacturing industry with low profits and weak competitiveness is vulnerable to the impacts of rising costs of raw materials and labor, as well as cyclical weakness in overseas markets.

In the past few years, Chongqing has innovatively developed large-scale processing trade. From the very beginning, it has paid more attention to the mode of industrial clusters. On the one hand, it has extended the industrial chain and kept most of the "smile curve" of processing trade in Chongqing. For example, 70% of the output value of each computer produced is from Chongqing. On the other hand, it has formed industrial clusters, especially the cluster of producer services and manufacturing industry, making the chassis of manufacturing industry more stable.

As a result, Chongqing has formed a world-class electronics industry base in a very short period of time. With the upgrading of the new energy vehicle and smart automobile industries in recent years, Chongqing, as one of the core cities of domestic automobile manufacturing, has again applied this experience of industrial clusters to automobile manufacturing.

According to the "2021 Industrial Added Value of Chongqing" released by the Chongqing Economic and Information Technology Commission on February 28, 2022, the output value of large-scale industrial enterprises has increased by 15.8% year-on-year, of which the automotive and motorcycle industry, the electronics industry, and the equipment industry increased by 11.3%, 17.3% and 15.9% respectively,

In addition, the proportions of these three industries in the city's large-scale industries are: 19.8%, 28% and 9.4% respectively.[1]

The automobile and motorcycle industry, especially automobile manufacturing, involves almost all known industrial categories, and is therefore known as the pearl of modern manufacturing. Electronics manufacturing involves almost the limit of human precision manufacturing, and is therefore known as the crown of modern manufacturing. Globally, the equipment manufacturing industry is known as the "Mother of all industries" due to its remarkable cornerstone status.

The automotive and motorcycle industry, electronics industry and equipment industry represent the advanced manufacturing capacity of a manufacturing city, and the proportion of these three industries in Chongqing is as high as nearly 60%, which fully demonstrates the achievements of Chongqing in the field of advanced manufacturing for many years.

This is also Chongqing's confidence of transformation from "manufacturing town" to "smart manufacturing town".

1　Zheng Sanbo, *Chongqing Daily*, "2021 Chongqing Large-Scale Industries: Enterprise Output Value Increased by 15.8%, with 1.998 Million Cars Produced", February 8, 2022.

Section II

Industrial Rebirth: From Urban Strategy to Corporate Rebirth

The year of 2001 was extremely special because China's two major applications in the international community were approved in that year.

On July 13, 2001, the International Olympic Committee announced Beijing's successful bid for the Olympic Games in Moscow, and the whole country was abuzz with excitement. On December 11 of that year, another news came from the shores of Lake Geneva, Switzerland. Although it also excited Chinese people, it did not receive as much national attention as Beijing's successful bid.

Twenty-one years later, we are certainly aware that the latter event was a unique event both in terms of its impact on China and the changes it made in the world, because on that day, a meeting at WTO headquarters approved China's accession.

This decision changed the global economic landscape for the next two decades. From that day on, the pace of global economic integration was accelerated, China quickly became the world's factory, and the Chinese economy was formally integrated into the global economy and played an increasingly important role as a new engine of growth in the process of global integration.

From Cost Performance to Disruptive Innovation

In the first decade of China's accession to the WTO, "Made in China" may not have taken the lead in disruptive innovation, but it has exerted its ability to control production and manufacturing costs to the extreme, which was the most core power source of China's transfor-

mation into the world's factory on a global scale. In the second decade of China's accession to the WTO, "Made in China" was not limited to playing the role of the world's factory. Thanks to China's huge consumer market and active innovation, a large number of Chinese manufacturing companies have begun to participate in global innovation and lead business change in many areas.

For more than a decade, China's smartphone growth has been almost a perfect model for Chinese innovation.

Over the past decade, in the smartphone market where major international giants are lined up, Chinese mobile phone manufacturers have collectively made a breakthrough in the past decade. In the midst of explosive growth of mobile Internet, Xiaomi has achieved great success by quickly integrating the latest technologies of the global mobile phone industry chain and winning the trust of consumers with the extreme cost performance. This competitive strategy quickly influenced Chinese mobile phone manufacturers and became the general tactics of the entire smart phone industry, thus forming a unique mobile phone industry pattern in the Chinese market. With the exception of iPhone, which can establish a long-term foothold in the Chinese market by virtue of its unique innovation ability, the top mobile phone brands have been completely replaced by Chinese brands. Moreover, Chinese brands even went abroad and successfully occupied a considerable proportion of the international market.

The importance of Chongqing manufacturing in the domestic market happens to be similar to that of "Made in China" in the global market in basic logic: the basic environment of manufacturing is relatively mature, labor costs are lower than coastal areas, and the speed of converting new technologies into new products is relatively fast. Therefore, within China's mature manufacturing system, Chongqing can easily replicate the success of "Made in China" in the global market.

Since the era of motorcycle manufacturing, Chongqing manufacturing is good at learning from the experience of Japanese and Korean motorcycle industry: its positioning is low-cost and high- efficiency, and new technologies that have been verified are manufactured into new products at a higher cost performance for consumers. In the era of automobile consumption, this competitive advantage is brought into full play.

In the summer of 2021, the writing team of this book visited Chang'an Automobile. What a technical manager said deeply impressed the team: the core competitiveness of Chang'an Automobile actually lay in bringing advanced technologies that have been verified in the market to consumers at a faster speed and at a lower cost.

In a sense, it is also the fundamental of Chongqing manufacturing in a long period of differentiated survival. This fundamental competitiveness is reflected in the westward migration of manufacturing industries such as laptop computers and mobile phones, as well as in the gathering of automobile manufacturers in Chongqing.

However, just as "Made in China" has to be upgraded to "Created in China" in the global market, Chongqing's manufacturing industry is also bound to undergo a transformation from cost performance to disruptive innovation.

Chang'an Automobile's Bottoming Out

In 2018, under the combined influence of various reasons such as the macroeconomic downturn, declining consumer spending power, high car parc and the previous overdraft of the purchase tax incentives, China's automobile market experienced negative sales growth for the first time after 28 years of rapid growth, falling 3% on the previous year, widening to 9.6% the following year.

Since then, China's automobile market has entered a "brand-new era".

It was in this context that Chongqing-based Chang'an Automobile experienced a cliff-like drop during this period. In 2017, Chang'an's cumulative sales reached 2.872 million units, down 6.23% on the previous year. In 2018, its cumulative sales were 2.1378 million units, down 25.6% on the previous year. In 2019, its annual sales were 1.759 million units, down 15.16% again on the previous year. In 2019, Chang'an Automobile lost as much as 2.647 billion yuan. It was the first annual loss in 23 years since its listing.[1]

---

1 *The Paper*, "Chang'an Automobile's Recovery Moment, Zhu Huarong Takes Over from Zhang Baolin as Chairman", June 11, 2020.

Chang'an Automobile's cliff-like drop shocked the whole auto manufacturing field in China, and aroused a deep thinking in Chongqing automobile field.

It is well known that Chongqing automobile was the "the third straight champion" in China before Chang'an Automobile's "cliff-like drop". In 2016, Chongqing's automobile production reached 3.16 million units, ranking first among all provinces and cities in China and accounting for over 11% of the national output, and was the first among all provinces and cities in China for three consecutive years. It was also the only city in China with an annual output of more than 3 million vehicles, making it a veritable "automobile city".[1]

As the core leading manufacturer of Chongqing automobile, Chang'an Automobile's cliff-like drop directly ended Chongqing's journey to the laurels.

Whether it is an automobile factory or an automobile city, it is an unachievable goal to recover quickly after suffering such a cliff-like drop. This means the long-term accumulation of the overall degradation of the competitiveness of a leading manufacturer, an industrial system, and an automobile city, or even the result of the superposition of three problems.

In this context, if we want to rebound from the bottom, in addition to finding the root of the problem from the enterprise and industry, and looking for new growth opportunities, we must also consider complex factors such as market changes, inertial brand gaps, and natural market response cycles.

The entire automotive field believed that Chang'an Automobile and Chongqing's status came to an end.

However, Chang'an Automobile shocked the whole industry again. After 2019, Chang'an Automobile ushered in a growth cycle with a new round of product debuts, and its sales bottomed out.

In 2020, Chang'an Automobile achieved annual vehicle sales of 2.004 million units, an increase of 14.0% on the previous year, breaking the 2 million unit mark again. In 2021, Chang'an Automobile contin-

1 Jiang Yan, Zhou Yali, *Chongqing Morning Post*, "Annual Production Capacity of 300,000 Vehicles: Chongqing Plant of Beijing Hyundai Completed", July 19, 2017.

ued to grow, with annual sales of its brands reaching 2.3 million units, an increase of 14.8% on the previous year. At the same time, Chang'an Automobile's operating income and net profit also rose steadily, turning losses into profits and continuing to make profits.

Up to now, the momentum of Chang'an Automobile has not disappeared. In March 2022, Chang'an Automobile's wholesale sales was 132,483 units, an increase of 20% on the previous year, surpassing FAW Volkswagen and SAIC Volkswagen, and once again rose to the first place among domestic manufacturers.

There must be the implementation of a rather critical transformation behind the strong momentum of Chang'an Automobile's bottoming out.

Continue to Internalize the Intelligent Dividend of the City

In 2017, Chang'an Automobile, which was aware of its own shortcomings (over-reliance on joint venture brands in sales and performance, and slow market growth), launched its third entrepreneurial plan, announced its transformation into an intelligent mobility technology company, and began to focus on "new energy", "brand up" and "intelligent".

At that time, the outside world believed that it was a risky move for Chang'an Automobile, after all, "the boat is too big to turn around", not to mention that the road ahead was full of reefs and unpredictable storms, which also added more potential risks to Chang'an Automobile. From 2017, in order to complete the transformation, Chang'an Automobile has successively launched its "Project Shangri-La" new energy strategy, "Beidou Tianshu Plan" intelligent strategy, and "No.1 Project" digital transformation project and the like, gradually improving the system capability. At the same time, the Blue Core powertrain based on the Blue Core NE platform has been applied successively, making Chang'an Automobile stand out in the market competition in recent years.

In addition, building a high-end new energy brand is a hot area where car companies are currently competing for layout.Whether it is the emerging of new car-making forces such as NIO, Li Auto, Xiaopeng

Motors, Evergrande Group, or the enhancement of domestic automobile groups such as VOYAH, ARCFOX and IM Motors since last year, they all have made the field increasingly popular. Of course, Chang'an Automobile, which is still in an important transition period, is not willing to be left behind.

In November 2020, Chang'an Automobile officially announced that it was working with Huawei and CATL to jointly build a high-end intelligent vehicle brand, including an intelligent electric vehicle platform, a series of intelligent vehicle products and a super "human-vehicle-home" smart life and smart energy ecology.

Enterprise innovation cannot be achieved without continuous investment in R&D, which is the material basis for enterprises to achieve innovation. In recent years, Chang'an Automobile has continued to invest 5% of its annual sales revenue in R&D. According to Chang'an Automobile's financial report, Chang'an Automobile invested 4.478 billion yuan in product and technology project development in 2019,[1] and R&D investment accounted for 6.34% of the operating revenue, which was in a relatively high level of the industry at that time.

In the future, Chang'an Automobile will continue to strengthen the construction of software capability in the field of intelligence. According to the plan, Chang'an Automobile will build an intelligent development team of 5,000 people in 2025, and the proportion of software developers will be increased to 60%.

From 2020 onwards, Chang'an Automobile has ushered in two major reversals: one is the widely known sales reversal, and the other is a deep introspective philosophy reversal.

Two facts can be summarized from this brief stumble in Chongqing automobile manufacturing.

One is the superb self-healing ability of Chongqing manufacturing. Even in the face of all kinds of impossibilities, it still achieves the rebound after the fall. The other is that the cost performance in the absence of innovation capacity contains great risks. When the demand of

---

1 Liu Yong, Yan Han, *Chongqing Daily*, "Chang'an Automobile Earns 600 Million Yuan in the First Quarter, Striving to Produce and Sell More Than 1.91 Million Cars This Year", April 30, 2020

the consumer market is changing, the technological innovation changes the rules of the game, and the development pace of an advanced manufacturing city is out of step with that of the country or even the world, which will cause severe consequences.

At the 2021 Smart China Expo, Zhu Huarong, chairman of Chang'an Automobile, said in his speech, "We have invested nearly 10 billion yuan in the past three years and now have mastered more than 500 smart low-carbon technologies, 48 of which are first launched in China. We have invested in 1.456 million intelligent connected vehicles, taking the worldwide lead."

Developing its own core competitiveness, leading advanced technological strength, upgrading Chongqing manufacturing in a new era, even in the long run, and realizing the intelligence and smartness of a city as a whole have obviously become the core tasks for the city.

Chang'an Automobile, together with many Chongqing smart manufacturing enterprises, is continuously internalizing the intelligent strategic dividend of a city into its own competitiveness.

Section III

Thinking Before Action: The Development Path of Chongqing Smart City

A city has its own life, and its own persistence and the fortunes of the times.

Looking back at the world's most famous cities, we always seem to see a small town like Macondo. A highlight moment of rapid evolution based on a special historical opportunity can often be found in the history of these cities.

Chongqing has not lost its manufacturing ability in different historical eras and several kinds of transformation, and finally has come to the era of the integration of industrial manufacturing and intelligent industry.

The Starting Point of a City of Dreams

Before 2017, while Chongqing's GDP was soaring, there was also a sluggish industrial layout.

It missed the opportunity for strategic transformation many times over a long period of time.

It did not get much coastal luck after China's entry into WTO as it was located in the western hinterland. Due to the lack of industrial base, it seized few opportunities in the wave of mobile Internet. It did not keep in line with the e-commerce boom that mostly driven by the supply chain of small commodities because of its heavy industry heritage.

Since 2017, Chongqing has been lucky enough to finally ushered in an opportunity to plan for its future from a more forward-looking perspective, and with more patience and more decisive practices. As a city famous for manufacturing, Chongqing's future opportunities can

only be augmented by AI for manufacturing. It should embrace the era of intelligence more boldly and arm itself with rich basic manufacturing resources via AI, achieving the key upgrade from traditional manufacturing to smart manufacturing.

On the surface, the difference between "manufacturing" and "smart manufacturing" is just one word, but in reality, there is a world of difference between them.

Because of such an opportunity, Chongqing feels the arrival of a new era, the accumulation of a new force, and the reorganization of a new life.

The intelligent era is coming. No one could have imagined that Chongqing would have the courage to put itself at the forefront of a cutting-edge field.

Countless AI bench-marking enterprises and innovation companies initially focused on cities such as Beijing, Shanghai, Guangzhou and Shenzhen, for the most cutting-edge AI technology needs to be matched with the freshest consulting, the best talents and the most active market. Moreover, when they had a closer look at Chongqing, a manufacturing city located in the west, they suddenly realized that all the ethereal and creative imaginations about smart manufacturing could be available here with the most direct and real manufacturing resources.

Through 100 years of industrial accumulation, Chongqing has become a nationwide industrial city with the richest manufacturing categories and the most complete upstream and downstream supply chains in China. In addition, Chongqing's manufacturing industries such as IT, electronics, software and equipment are also leading in the country in recent years, and it is also one of the five top nodes of the national industrial Internet identification analysis. All of these form a mature industrial Internet environment that is enough to attract any enterprise and team.

Thus, since 2018, countless stories have been put on the stage and in booths of the Smart China Expo.

Numerous elites in the field of intelligence walked off the stage of the Smart China Expo and couldn't wait to start to look for offices in the skyscrapers. Countless AI enterprises walked out from their booths at the Expo and couldn't wait to find partners in the factories.

Behind these elites and enterprises, there are a number of enterpris-

es that meet in Chongqing by chance and will inevitably take root here, as well as a number of opportunities to resonate with the world and rise with the trend of the world.

Perhaps with the hindsight of Chongqing locals, Chongqing has already become a global destination for intelligent industries.

Sometimes, the city's key choices can leave a significant mark on history. Compared to these moments, the hardships of the past are not so important.

Chongqing is releasing its manufacturing vitality from the factory floor and enhancing it in booths, on lecture stages, and in the digital world. The Smart China Expo is an important transformation opportunity for Chongqing.

In contemporary times, the long-term vitality of an enterprise, an industry or a city is no longer measured by profits, revenues and GDP, but by the use of technology to activate the industrial chassis of smart manufacturing, which in turn continuously outputs future-oriented competitiveness.

Chongqing's Imagination in the Era of Intelligent Industry

In the morning mist with mountains and rivers, in the night with fresh fragrance and spicy food, people from all over the world are labeling the city of Chongqing with new labels one after another with their own perceptions.

In today's Chongqing, half of the labels are related to intelligence.

After nearly 100 years of manufacturing accumulation, Chongqing has finally been supported by AI with big data and intelligence as the base. The combination of the hard-core skeleton and the soft blood immediately creates an urban vitality in the digital age. The development of Chongqing has becomemuch clearer: the goal is to build a "smart manufacturing town" that is featured with blood, flesh, vitality and ecology, which is also a "smart city" that is greener, more beautiful, and more livable.

For a long time, about what a smart city is, in fact, each city has its own different understandings, interpretations, visions and practices. For

Chongqing, which is determined to build a "smart manufacturing town" and a "smart city", a clear and complete smart city is on the horizon.

In short, "smart manufacturing town" and "smart city" are two aspects of Chongqing's smart city strategy, which are to build the future pattern of Chongqing's smart city from the perspectives of urban economy and urban life.

If we follow the thinking to review the steps of smart city advancement in Chongqing, it is not difficult to find scientific and practical aspects. To put it simply, it can be divided into three steps from the perspectives of economic development, governance structure, and social life. We will build the entire industry chain of "chips, LCD, smart terminals, core components and IoT", the all- element cluster of "cloud, Internet, data, algorithm and application", and a full-scenario set of "housing, education and employment, tourism, entertainment and shopping", gradually promoting the digitization of economic structure, infrastructure, and life applications, with a clear strategic level and a reasonable pace of advancement, which is more in line with the essential pursuit of smart cities.

The Smart China Expo plays a particularly important role in the whole process of strategic advancement. It is not only a window to look at the world, but also a door to embrace the future. With the gradual advancement of Chongqing's smart city strategy at different stages, the continuous hosting of the Smart China Expo will keep providing key support.

The year of 2022 marks the fifth year for Chongqing to hold the Smart China Expo, as well as the fifth year of Chongqing's deep docking with a cutting-edge industry. As soon as the gate of the Smart China Expo is opened, the news spreads and the industry comes in. With more and more innovative enterprises with unlimited possibilities in niche areas setting up their AI headquarters in Chongqing, the city's smart manufacturing industry ecology has begun to take shape.

So far, with the hard-core skeleton and the soft blood, Chongqing has integrated the bone and blood of the intelligent industry. Chongqing's smart manufacturing gene has also completed its own mutation in line with the times. There is a bright future ahead. Although there will be difficulties in self-transformation, there will be no confusion about where to go.

Smart City: What Is the Future of Chongqing?

In the past, people used to evaluate the development level of a city by indicators such as economy, governance, cultural environment and average revenue. The indicators themselves will continue to exist to evaluate the status of cities regionally, nationally and globally.

In the intelligent era, exploring the development of smart city will become a new common goal for most cities. How to use the new generation of information technology to realize the following goals has become the common mission of "smart city" construction for global cities: achieving the deep integration of information technology, industrialization and urbanization, alleviating the "urban diseases", improving the quality of urbanization, realizing fine and dynamic management, enhancing the effectiveness of urban management, and improving the quality of life of citizens.

We've talked a lot about Chongqing in the present, so what will the future hold?

The Future of Smart Cities Is Not Only Intelligent Cities

Many people may wonder how many cities there are on this planet.

Obviously, due to the natural differences in economic development, settlement habits, administrative divisions and urban standards of different countries, the definition of a city varies from country to country, and the answer to this question is even difficult to find by conventional statistical methods.

With such curiosity, a team led by Wu Zhiqiang, an academician at

Tongji University, began to use AI to solve the question in 2014, and finally got the answer in January 2018: there are 13,810 cities and built-up areas with an area of more than 1 square kilometer in the world.[1]

Wu's team not only counted all the cities in the world, but also drew the edges of the curves of these 13,810 city trees precisely. Since 2014, Wu and his team have developed the "city tree" global image intelligent identification technology, which is to identify and superimpose the satellite remote sensing images of all cities in the world within a 40-year time span through a 30m×30m precision grid. It is named "city tree" because the spatial-temporal evolution of the city is visualized in a tree-like trajectory.

For these 13,810 cities counted for the first time in the world, Wu's team even established a world city evolution database using AI technology. They used AI image recognition technology to find the scope of urban built-up areas, to sum up the development laws of world cities, and even to deduce the development scenarios of cities in 2020, 2035 and 2050, to diagnose the future problems of cities, and to provide support for the study of human city problems in the whole world.

As human society enters the intelligent era, 13,810 mayors around the world are facing a question of the times: how to build their city into a smart city?

The visions of smart cities in the minds of 13,810 mayors may be different, but it is not difficult to summarize some of the consensus that has been formed during the global exploration of smart cities in the past decade.

Smart cities are definitely more than intelligent cities.

Smart cities can improve the efficiency and convenience for economic development, urban governance and citizens' lives by using various information technologies and intelligent applications, but they cannot stop there. Smart cities need the support of AI technology, but they cannot let the fate of the whole city to be controlled by the machines. If we simply automate all kinds of IoT devices, set them up and run

---

1 Zhang Jiongqiang, *Xinmin Evening News*, "Tongji University Wu Zhiqiang's Team Draws the World's First Complete Map of 13,810 Urban Built-up Areas", August 12, 2019

according to AI programs, and embed various instruments in all corners of the city, then the city can only be reduced to a huge lifeless machine with the appearance rather than the essence of a smart city.

The participation of human intelligence is even more important for smart cities.

The reason why it is called "smart city" instead of "intelligent city" is that talents are the core of the whole smart city. Cities serve the people who live in them, and people are rooted in fully-technological cities. AI technology can simplify those cumbersome, mechanical, inefficient and costly urban operation modes, and the remaining space should be left for people to live their personal lives and think independently.

Smart city, Focusing on Open Innovation Ecology

For a long time, the focus of smart city strategy in many cities often falls on the new generation of information technologies represented by IoT, cloud computing and mobile Internet, blindly pursuing various smart city projects while ignoring the development of smart industry in cities. For such smart cities, even if they are built, it is difficult to form their own "city life" and have the ability to evolve for the future.

Compared with the intelligent technology hardware of smart cities, the software that is gradually nurtured in the knowledge society environment is actually more important, i.e., the innovative enterprises and innovative talents who continuously contribute their wisdom to the iteration of smart cities, and the nurturing environment created for them by the whole city.

Smart cities should have their own open innovation ecology.

Therefore, the final form of the smart city is not a group of buildings and a bunch of hardware. The development of the smart city not only relies on the progress of various technologies, but also needs to be promoted by the development of the society, forming an evolving and growing model.

At the same time, the introduction and rooting of innovative enterprises require fertile soil for an innovative economy; the absorption and growth of innovative talents also require abundant illumination from

innovative industries. When the intelligent era is approaching, Chongqing will resolutely move towards intelligent manufacturing, laying a key foundation for economic development, and will continue to provide innovative impetus for the construction of a smart city.

Take the integrated circuit (IC) design industry as an example. In 2020, the sales growth rate of IC design industry in Chongqing reached 206.1%, with three enterprises achieving sales of more than 100 million yuan, and the year-on-year growth rate of "chips" and "LCD panel" exceeded 20%.[1]

Behind the rapid development of the IC design industry is the interaction and empowerment between different industries. Chongqing has a market demand of 200 million intelligent terminals, 60 million notebooks, 30 million home appliances and 2 million cars. The entire industrial chain of "chips, LCD, smart terminals, core components, and IoT", the all-elements cluster of "cloud, Internet, big data, algorithm and application", and the full-scenario set of "housing, education and employment, tourism, entertainment and shopping" need more diversified support of semiconductor technology, standards and products. The semiconductor industry has become the industry with the widest application and deepest penetration in Chongqing's industrial chain. CloudWalk Technology Co., Ltd., known as one of the AI four little dragons in China, was also born here.

Chongqing's achievements made in the past five years will eventually go down in history.

In the past five years, Chongqing's GDP has increased by nearly 1 trillion yuan with an average annual growth rate of 6.7%, entering the 2 trillion club in 2018, reaching 2.79 trillion yuan in 2021, with the per capita regional GDP reaching 87,000 yuan, higher than the national average, and it is expected to reach a new record in 2022. Supply-side structural reform continued to deepen, industrial transformation and upgrading accelerated, and pillar industries such as automobiles and electronic information continued to grow. Innovation-driven development led by big data intelligence had been further advanced, and break-

---

1 Zheng Sanbo, *Chongqing Economic Times*, "IC Design Industry: 206.1% Growth in Chongqing Last Year", January 18, 2021.

throughs had been made in the construction of "smart manufacturing town" and "smart city". Four Smart China Expo had been successfully held, and the added value of digital economy accounted for 27.2% of the regional GDP. Chongqing seized the opportunities from the national level. Many achievements such as the joint planning of Chengdu-Chongqing twin-city economic circle, the introduction of a number of important planning programs, the successful implementation of the landmark major projects of Chengdu-Chongqing high-speed railway, the accelerated construction of regional cooperation platforms, such as Wanzhou, Dazhou, Kaizhou, Sichuan and Chongqing Integrated Development Demonstration Zone, and the approval of the establishment of Chengdu-Chongqing Financial Court, have already laid a solid foundation for building the fourth pole of China's regional economy in the future.

Tracing the origin of these achievements, it is not difficult to find the key node five years ago: a city of heavy industry located in western China firmly pushed open the door of the intelligent era when a new era came, and bravely marched forward.

Five years later, the whole world find that the intelligent era has a profound impact on Chongqing. Whether it is the annual Smart China Expo or highly automated smart factories, as well as the confidence of the citizens scattered in the city, the talk in the hot pot restaurants, and the discussion in the conference room, are the distinct marks of the intelligent era.

In the next five years or so, what will happen to Chongqing, which is already deeply bound to the intelligent era? It is already a matter of close concern to all who care about the city's future.

During the 2021 Smart China Expo held from August 24 to 25, Lijia Smart Park launched "A Day of Smart Life", providing a space for visitors to experience the future of smart life in advance.

The application scenario of "A Day of Smart Life" is a large-scale activity project which integrates the themes of smart living, smart life, smart work, smart medical care, and smart arts and interests created by Lijia Smart Park according to the logic of spatial transformation of daily activities and the standards of immersive and authentic feeling. It includes 60 experience scenarios and 130 experience projects, fully

demonstrating the leading role of intelligence in high-quality life.

The 130 experience projects of "A Day of Smart Life" are fully human-centered. It applies the most advanced IoT and big data technology to build a smart life full of technological intelligence and human insight.

So, is this what the future holds for Chongqing? Perhaps, no. Information technology, data algorithms and hardware innovation have always created new intelligent industrial species in ways that have overturned human imagination in the past decades. Our imagination today is limited to the way we already understand and does not fully represent the future. Beyond the edge of human imagination about the future, there is a richer space for AI innovation, and on the day when the real smart city is presented to us, there will be endless surprises.

But for Chongqing, this is already a good start. Being in the era of intelligent industry, gradually building the city's digital economy, digital governance and digital life will surely accumulate more and more experiences and resources.

The future of Chongqing is built on all these explorations and is worth imagining, worth being proud of and worth believing in.

Afterwords

In order to echo the new trend of the development of the intelligent era, to show the new achievements of the intelligent industry, and to create a series of excellent popular science books for the general public to understand and integrate into the intelligent era, the Publicity Department of the CPC Chongqing Municipal Committee decided to publish the *Decrypting the Intelligent Era Series* on an ongoing basis in 2020 and requested that the *Decrypting the Intelligent Era Series* be made a cultural card of the Smart China Expo with international standards.

The two books published this time are the continuation of "Decrypting the Intelligent Era", a "cultural card" born with the Smart China Expo. The Series is coordinated and organized by the Publicity Department of the CPC Chongqing Municipal Committee. Jiang Hui, Member of the Standing Committee of the CPC Chongqing Municipal Committee and Director of Publicity, instructed that it should focus on the overall strategy of building an "intelligent manufacturing city" and a famous "smart city"in Chongqing, around the five-year Smart China Expo and the achievements of the intelligent industry, and do a good job in the publication of this year's *Decrypting the Intelligence Era Series*.

There are two books in the Series this year, including *Decrypting the Intelligent Era: Overlook the Global Intelligent Industry from the International Smart China Expo (2018—2022)* (bilingual) in the form of illustrations, which combs the development of the global intelligent industry, shows the essence of the previous Smart China Expos, reviews the achievements of Chongqing's smart city construction, and also analyzes and foresees the future development of Chongqing in the era of intelligent industry and the future practice of smart city construction; and

Decrypting the Intelligent Era: Redefining the Smart City, which shows how big data intelligence can promote organic urban renewal, actively expand the characteristic smart application scenarios, and explores the practical path of an international, green, intelligent, and human-cultural beautiful city.

During the whole writing and publishing process, Cao Qingyao, Executive Vice Director of the Publicity Department of the CPC Chongqing Municipal Party Committee, gave comprehensive guidance to the Series, and the Publishing Department coordinated several professional teams to work closely and made an overall review of the planning and creativity of the Series as well as the quality of the contents to promote the completion of the preparation and publication of the book. The research, writing and publication of the Series were also supported by Chongqing Municipal Commission of Economic and Information Technology, Chongqing Municipal Bureau of Big Data and the Secretariat of the Smart China Expo.

Chongqing University Press invited Director He Yonghong of the Secretariat of the Smart China Expo, Professor Xiang Hong of the School of Big Data and Software of Chongqing University and Professor Zeng Deheng of the School of Management Science and Real Estate of Chongqing University to review the Series, and Rao Banghua, Director of Chongqing University Press organized a number of senior editors to revise the manuscript word by word to ensure the scientificity, readability and accuracy of the content.

If we stand on the dimension of the historical process of the intelligent era, we hope that the *Decrypting the Intelligent Era Series* can

record and look forward to how intelligence actually empowers the economy and adds color to life, record and look forward to the practical process of "digital industrialization and industrial digitization", record and look forward to this great and profound change in the history of human civilization. Such a record and outlook, such an annual imprint of the intelligent era, is of historical significance.

Hereby, we would like to pay tribute to Smart China Expo, and express our deep gratitude to all the organizations and individuals who contributed to the project establishment of this book, provided writing materials, wrote and translated the book, participated in the review of this book, helped the publication of this book, and to the experts interviewed by the writing team.

Writing Team
July, 2022